Resilience Through Cyber-Informed Engineering:

An Engineering and Operations Approach to Cybersecurity

Resilience Through Cyber-Informed Engineering:

An Engineering and Operations Approach to Cybersecurity

Andrew Ohrt, PE, CISSP
Daniel Groves, PE, CISSP

American Water Works Association

ISBN: 978-1-64717-153-7   ISBN, electronic: 978-1-61300-768-6
https://doi.org/10.12999/AWWA.20867ed1

Sr. Manager — Product Acquisition & Development: Geoffrey S. Shideler
Manager — Publishing Operations: Gillian Wink
Specialist — Copyright and Permissions: Peggy Tyler
Director of Publishing: John Fedor
Technical Editor: Wiley
Technical Editing Review: Suzanne Snyder
Cover Design and Technical Illustrations: Michael Labruyere
Cover Images: Rustic, klyaksun, Vegorus, StockSmartStart/Shutterstock.com
Production: Innodata

Library of Congress Cataloging-in-Publication Data

CIP data pending from the Library of Congress

American Water Works Association

Contents

List of Figures

List of Tables

Dedication

To our families.

To all of water and wastewater professionals who provide their communities safe and continuous service.

Note to the Reader

With all new technical references, there has to be some new terminology. Most important in this book are

- cyber-informed engineering (CIE);
- consequence-driven, cyber-informed engineering (CCE); and
- critical function assurance (CFA).

All three of these approaches/terms/frameworks and many others are discussed in detail through this book. In application, organization, and implementation, they are very different.

However, they were all born at Idaho National Laboratory, by many of the same people with the same end in mind. Arguments are had among practitioners about which one is best for a given situation. Practitioners have those types of arguments. At the end of the day, it doesn't really matter to most of the world. The objective of each as applied in our sector is the same:

Ensure our water systems are resilient to sophisticated cyberattacks today and in the future.

To simplify this text, in many cases the authors use CIE as an umbrella term for all three approaches. The exception is if one approach is used or called out specifically.

There are many resources supporting each of these concepts. Your favorite search engine is ready to help you find and explore those. We recommend using the term that is most useful to you. For us, as engineers, CIE is that term.

Andrew Ohrt & Dan Groves

Foreword by Andy Bochman

Seeing that everyone around you is heading over a cliff is more than a bit unsettling. But realizing that, despite your warnings, they are determined to continue full speed in that direction is a simultaneously humbling, frustrating, and frightening feeling. It's one thing if you're a lemming. It's something altogether different if you're a professional engineer designing systems that undergird civilization. Those systems provide health and safety benefits we enjoy but also take for granted. And in the wrong hands, they could bring disease or death to hundreds, thousands, or more.

AN ENGINEER TAKES SHAPE

Born in the 1970s while Mississippi was still trying to figure out how to get desegregation finished, young Dan Groves was no stranger to challenges. However, he was drawn to the concept of engineering and designing things while working with his father building homes at age 5. Dan was also naturally drawn to taking things apart to figure out how they worked, sometimes fixing broken appliances and gadgets from TVs to coffee makers and photographic equipment, simply through trial and error. Talk about having an engineering mindset from early on!

TWO WHO SENSED DANGER EARLY

In the early 2000s, only a few electrical engineers were thinking about cybersecurity, if any at all, and only a handful understood what the widespread move to internet protocol (IP) ominously portended for their discipline. Daniel Groves, now Business Sector Lead at West Yost, an advanced water sector engineering firm based in Davis, Calif., was one of them. Another was Navy intelligence officer Mike Assante. The two never met in person, but in an example of long-distance synchronicity, they came to similar conclusions about the urgency of the problem and what might be done to counter it.

In the mid-to-late 2000s at the Idaho National Laboratory (INL), the as-yet-unnamed concept for protecting critical industrial control systems

(ICS) was crystallizing into what Mike simply called "the Framework." Its nascent form was captured on paper and pixels in an image INL colleagues called "Curtis's brain," a reference to Mike's friend and collaborator, veteran INL engineer Curtis St. Michael, and his mind-boggling, multidirectional workflow diagram that really did capture the content and intent of the Framework.

TROUBLE LOOMING BENEATH THE SURFACE

Following instructions from the control systems manufacturers who made the products he configured, among his early experiences in the field was designing PLC*-based water management systems with proprietary communications protocols. At the time, these systems were not connected in any way to business networks, so much so that the now-discredited cybersecurity term "air gap" honestly seemed to apply. Good thing, too, as cybersecurity issues were emerging and rapidly multiplying in business IT systems and networks.

Then in 2001, for the first time, Dan had to design an ICS using Ethernet (another term for IP) communications, and thus began the transformation and journey he's still on today. Realizing the world was about to change in important and dangerous ways, his was a lonely position surrounded by otherwise skilled engineer colleagues who didn't seem to understand what was happening. Responding to his calls for caution with phrases like "It'll never happen," "It's no big deal," and "You're making way too much of this," they were all but calling him Chicken Little.

He had been trained that infrastructure engineers err on the side of caution—for example, building a bridge able to carry more weight than the estimated heaviest load it would experience in its projected life span. Dan thought that had to apply to security too, as access to control systems could allow a cyberattacker to cause grievous damage to the water and wastewater treatment systems folks like him were paid to design. That was Dan's mindset, at least.

Then came a moment when everything crystallized.

Having recently been given a copy of *TCP/IP for Dummies* by a friend, at the same time he had on his desk a PLC, an Ethernet switch, and an uneasy feeling—it dawned on Dan that the PLC, a device used to control physical processes (a.k.a. machines), was now on the business network, and so were many of the systems he had designed that were currently in operation managing water and wastewater facilities across the country. He dug in deep and completed a Microsoft (MS) Certified Systems Engineer certification. Don't

---

* PLC stands for programmable logic circuit, essentially a small computer that can be customized to direct physical devices (e.g., actuators, pumps, valves) to perform their tasks (or not) under predetermined sets of conditions.

be fooled by the word "engineer" in that title; this was all about how the MS operating system and IP networks can connect anything to almost anything.

So he went to town.

As part of ongoing self-study, he was introduced to a tool called Wireshark. Back at the office, Wireshark allowed him to scan the corporate network, and he saw his boss's e-mail and password go by in plain text. Shocking, yes, but all the more so because he learned to do it in just 5 min with the tool. If someone with almost no experience could do something this powerful so quickly, imagine what a skilled user could do. It was a terrifying revelation. It followed that determined attackers, then, could do whatever they wanted; they could always get in and just about no one would know it. As former CIA director General Michael Hayden warned, "They're going to get in . . . get over it!"

Something had to change.

But first, he had to endure more disbelief and ridicule. Colleagues at Malcolm Pirnie—his company before West Yost—told him he should "chill out." "You are overreacting." "You are just fear-mongering." "What we're designing is isolated, air gapped." A sneaking feeling crept in: "Could I be wrong?" When so many supersmart senior engineers were confident there was no problem, wasn't it possible, if not likely, that there was *no problem*?

About those confident engineers.

All during his training and early days on the job, Dan observed that engineers strive to use only proven and well-known tools and techniques. They rely on authority and precedent and make copious citations to standards and codes in every specification. Maybe if Dan could get a standards body to say that cybersecurity for ICS was important, engineers would start to act accordingly, and things would start to improve. He searched and couldn't find any standards body (e.g., NCEES, NIST, IEEE, ISA, IEC) even mentioning cybersecurity in passing. So in 2011 when NIST 800-82, *Guide to Industrial Control Systems (ICS) Security*, arrived, it was a very big deal. Finally, an authority that engineers respected was recognizing this was an issue for control systems.

SEA CHANGE

Years passed and several revisions of NIST 800-82 witnessed engineering communities still slow to awake to the core issues. Dan was struggling to best communicate the urgent ideas he had formulated in 2018 when he happened upon a *Harvard Business Review* article provocatively titled "The End of Cybersecurity." It was the first public introduction of consequence-driven, cyber-informed engineering (CCE), formerly known as the Framework, and the cyber-informed engineering part was what resonated most with him. A thought flashed: This methodology, coming out of a US Department of Energy national laboratory renowned for its applied engineering and control systems leadership, has the potential to change the way all engineers think.

2025 finds West Yost licensed and trained by and partnered with INL. And with Dan's buoyant and clearly visionary leadership, West Yost brings its own water sector–specific, tailored CCE training sessions and engagements to clients. This year also sees the company, arm-in-arm with AWWA, publishing its own book for water sector defenders with customized case studies, all targeted and relevant to water infrastructure defenders.

Dan shared with me recently that his aim is to "bend the arc of cyber" in the water sector. Mike, who passed too young a few years ago, sure would have liked that.

Preface

In February 2020, a stone's throw from the frozen Snake River, an idea was born that could change traditional engineering practices. Staff from Idaho National Laboratory (INL), AWWA, and West Yost Associates (West Yost) convened to discuss how to bring consequence-driven, cyber-informed engineering (CCE) to the water sector. Presented to the world in Andy Bochman's 2018 *Harvard Business Review* (*HBR*) article, "The End of Cybersecurity," CCE is a return to engineering first principles, dovetailed with digital technology as a way to manage current and future cyber risk.

Whenever Dan introduces himself, he speaks of his cyber awakening that occurred in the late 2000s. For some time before that, he was already a control systems engineer. One day he realized that the control systems that utilities were installing were fundamentally lacking cybersecurity as a key design element and thus were very insecure. This resulted in his grassroots effort to convince clients and other practitioners that the vulnerable designs were a problem. By his own admission, many engineers in the water/wastewater sector were not of the same mindset and ready to agree. However, after reading Andy Bochman's May 2018 article, "The End of Cybersecurity," published in the *HBR*, he knew he finally had the tool to convince people his thinking to date was correct. Andy's article in the respected and widely read publication *HBR*, coupled with the fact that it was written by an INL staff member, did start to turn heads in the water sector. After several attempts to connect with Andy and INL, Dan was finally successful in late 2018. Several of us had the chance to meet in December 2018 with Andy in Minneapolis, Minn., and forge the beginnings of a relationship that has been both professionally and personally rewarding.

My path to CCE was a little different. I first saw Andy speak at the University of Minnesota's Technological Leadership Institute in 2017. At the time, I was very focused on all-hazards resilience, a discipline of which cybersecurity was a small but growing sliver. Other than being impressed by the guy from INL who was speaking about cybersecurity in a manner that I had never heard before, I have to admit I didn't retain those three magical letters: C-C-E.

Fast-forward to riding the elevator with Andy to that same meeting in December 2018. As I sat through that meeting, it was one of those strange moments that has occurred only a handful of times in my life. I knew then what I needed to go and do next —bring CCE to the water sector.

CCE was developed by a talented group of INL engineers and cyber-security experts. This group had diverse backgrounds in nuclear and electrical control system engineering, the military, offensive cyber-warfare operations, commercial cybersecurity, and more. When applied to the water sector, CCE becomes a critical approach to managing cyber risk.

It is not frequent enough that utilities and their service providers have the cybersecurity knowledge and resources to manage this risk. However, as a sector, we have a long tradition of excellent engineering practices that date back thousands of years that ensure the delivery of safe drinking water and collection of wastewater to protect our customers, communities, and the environment. Much of our engineering is implemented in the physical world, despite the inevitable digitization and automation. CCE provides the bridge, recognizing the value of and inevitability of automation while allowing us to maintain the security of our infrastructure, people, and service to our customers.

When we conceived of this book and agreed with AWWA that it should be published, it was 2020. The expected publish date is October 2025. A five-year delay is quite long. An argument could be made that there is an opportunity cost to our delay. With all of the infrastructure that was designed, built, and commissioned in that period of time without this book as a resource, that cost is real for a portion of our sector.

However, as we consider the bigger picture, many things have changed. Our threat environment continues to evolve while we have worked with our clients to implement the concepts described in this book. All best practices take some time to sort out. These are no different. We have experimented and found what was most helpful to our clients and, ultimately, to the sector. This is a much more complete book now than it would have been in 2020.

—Andrew Ohrt, Duluth, Minn.

*   *   *

Primary Authors

Andrew Ohrt, PE, CISSP, is the Resilience Practice Area Lead with West Yost and a proud member of the Beer-ISAC. He has supported more than 60 utilities of all sizes to build all-hazards resilience. Andrew is based in Duluth, Minn.

Daniel Groves, PE, CISSP, is the Operations Technology, Cybersecurity, and Resilience (OTCR) Business Sector Leader with West Yost. With more than 25 years of experience, Dan was a journeyman electrician before becoming a *licensed* electrical engineer and control systems engineer. Dan is based in Phoenix, Ariz.

Who Is This Book For?

This book is intended for the "One Water" sector. The One Water concept has been around for many years and is increasing in relevance. Broadly, this can be defined as the use of water from a source, discharge of treated water, recycling of that water in the natural environment, and everything in between. One Water encompasses a wide range of organizations and people in a variety of positions from entry-level operators and technicians to chief engineers and general managers with decades of experience. This includes the variety of professionals, regulators, and professional organizations within and in support of utilities and their mission. Anyone who operates an engineered system within the broad category of One Water should consider this book relevant to their operations, engineering, management, and decision-making.

- Leadership—general manager, CEO, director, city manager, board/council member
  - People in these positions hold fiduciary responsibility for the sound management of their organizations on behalf of citizens, shareholders, and customers. This responsibility now extends to cybersecurity and cyber resilience of their systems.
- Engineering staff—civil, water resources, electrical, mechanical, control system engineers and planners
  - Our professional engineering standard of care is changing. Our understanding of cyber risk continues to evolve, and with it, the engineering practices to address this risk.
- Operations staff
  - Engineers must be held responsible to design a system that can function under an attack/loss of automation.
  - Operators must be prepared to recognize the precursors of a cyber-attack and response to ensure operational continuity.
- Emergency managers
  - These ideas resonate with emergency managers because of focus on engineering redundancy and ensuring staffing capabilities.
  - Cyber-informed engineering (CIE) emphasizes response and recovery planning, training, and exercising.

- o Risk managers
  - Reduce the need for risk-transfer mechanisms, thus reducing costs and risks to insurers (if these risk-transfer mechanisms are even available).
- o Finance managers
  - Bond rating agencies are adapting their ratings systems to include cybersecurity.
  - Implementation of CIE may support improvement and maintenance of bond ratings.
  - If a utility has implemented CIE within the organization, in the opinion of the authors, this demonstrates a mature approach to cybersecurity and worthy of note when the utility needs to issue bonds.
- o Insurers
  - Insurers are some of the most sophisticated risk professionals in the world. Improved engineering practices will reduce their cyber-risk exposure beyond what can be done with standard cyber-hygiene practices.
- o Regulators
  - Regulatory agencies are, by and large, mandated with ensuring drinking water is safe, human health is protected, and ecosystems remain healthy.
  - Regulators must recognize that new monitoring requirements may introduce new technologies and thus vulnerabilities.
- o Original equipment manufacturers (e.g., SCADA hardware) and systems integrators
  - Manufacturers and systems integrators must integrate improved cybersecurity approaches into their hardware and software provided to the water sector.
- o Capital planners
  - CIE must be considered when implementing new infrastructure.
  - Improved risk management within the planning and implementation of new infrastructure includes CIE.
- o Information Technology (IT)
  - Many journal articles, blog posts, and conference presentations have documented the phenomena of IT/OT convergence. In many ways, this began over a decade ago. Traditional IT approaches have not been wholly successful in eliminating the consequences of cyber-attacks in a manner that CIE has the potential to be.
- o Health and safety
  - Ensuring the health and safety of staff members, contractors, and the public is one of the most important efforts undertaken by utilities today. Implementation of these principles emphasizes the protection of people from hazardous exposures and dangerous working conditions.

- o Procurement
    - Improving the procurement of hardware, software, and engineering services in support of engineering and operations will set water systems up for improved cyber resilience.
    - Prevent information from leaking into public forums.
- o Professional Engineering Boards and advocacy organizations such as National Council of Examiners for Engineering and Surveying (NCEES)/National Society of Professional Engineering Boards (NSPE)/ American Council of Engineering Companies (ACEC)
    - CIE enhances the safeguarding of health, safety, and welfare of the public with respect to current and future cyber-risks to critical infrastructure.

How This Book Is Organized

This book is organized generally in the way the authors developed an understanding of this area of practice.

Chapter 1—First, we need to understand how we got where we are from a cybersecurity risk to critical infrastructure perspective.

Chapter 2—Why and how is CIE relevant to the water sector, and why is it different from how we have addressed cybersecurity?

Chapter 3—To understand the importance of and be ready for CIE, we need to know what "A Day Without SCADA" would look like and its effect on the sector.

Chapter 4—Many in the water sector have been practicing the concepts of CCE/Critical Function Assurance (CFA) for years. But how do we maintain these practices as technology continues to grow and evolve in the sector?

Chapter 5—CIE is best understood via examples. This chapter provides several real-world case studies to help.

Chapter 6—Now that you understand CCE/CIE/CFA, what do we practically do now? Technology is continuously changing. How can CCE/CIE/CFA help us address emerging technology challenges?

Acknowledgments

First, we would like to thank Andy Bochman and Virginia "Ginger" Wright. Collectively they have been like our Yoda, Galadriel, Gandalf, Morpheus, Dumbledore, and Obi-Wan all wrapped together. Without their advocacy, support, insights, guidance, and patience, this book would still be "something that we should really do."

Thank you to Dr. Kevin Morley from AWWA for being in lockstep with us on this topic since that first meeting in Idaho Falls in February 2020. The water sector is better off with you looking out for us in Washington.

Numerous water and wastewater utilities informed the case studies presented in this book. They shall remain anonymous. Please continue to champion these practices within your organizations.

Thank you to all of the additional past and current Idaho National Laboratory staff we have worked with over the last five years, including: Amanda Belloff, Micah Stephenson, Rob Smith, Jon Cook, Nathan Johnson, Curtis St. Michael, Dave Kuipers, Ben Lampe, and Jessica Quick. They have been wonderful to work with and have been the source of many unique experiences and fond memories for us.

We have benefited from many conversations with friends across sectors, including Arlen Busch, David White, Monica Tigleanu, Patrick Miller, Jerry Cavaluzzi, Sarah Walsh, and David Yonge. Not only did we learn much from these individuals, but they also cheered us on throughout the process.

Publishing a book, like delivering any long-term project intending to change the world, is a team effort with many people working on tasks foreign to us. Thank you to the entire AWWA publishing team. Most prominently, Geoff Shideler, who gave us the right push at the right time to finally cross the finish line.

Thank you to all of our colleagues at West Yost. Thank you to Derrick Bouchard, Amanda Jones, and Derek Zohner for help with graphics. Thank you to our colleagues—Joel Cox, Jeff Hesse, Greg Smith, Tara Mertz, Abigail Madrone, David Garrison, Zane Wilsterman, Jeremy Smith, Kenny Smith, Jeff Pelz, Greg Chung, and Michael Gruenbaum—for being sounding boards and grinding it out with us day to day. You no longer have to ask "so, how is that book coming?"

The list presented above is inevitably incomplete. To anyone who we may have missed, we are thankful for your contribution.

Finally, thank you to you, the reader. We hope this work is helpful, your organization is more resilient, and you sleep more soundly because of implementing the concepts discussed.

Chapter 1

How Did We Get Here?

Your first question might be, *Where is here?* If that is the case, welcome! "Here," to answer the question, is our current situation, where many of our water and wastewater systems cannot be expected to be resilient to a cyberattack. We have arrived at this point because, as humans, we haven't adequately characterized the risk associated with digitizing and networking all of the physical assets that are necessary to pump, treat, deliver, and collect water and wastewater. Over the last three decades, engineers, operators, vendors, consultants, and integrators have made increasingly detrimental assumptions about the security and resilience of the digital systems that their organizations and clients have come to rely on. This book and cyber-informed engineering (CIE) are intended to unwind that resilience debt. Resilience debt is a more common term used in climate change resilience circles, but generally, it means the debt accrued through engineering assumptions that do not adequately prepare a system for disruptions (Bochman 2024).

WHAT IS CIE ANYWAY?

In January 2021, Andy Bochman and Sarah Freeman published *Countering Cyber Sabotage: Introducing Consequence-Driven, Cyber-Informed Engineering (CCE)*, ISBN: 978-0-367-49115-4 e (the "CCE Book"). The authors recommend that any reader who would like to build a deeper understanding of consequence-driven, cyber-informed engineering (CCE) and how to implement CCE within their organization should purchase and read the CCE Book.

Andy Bochman once told me that CIE can be likened to the tree, and CCE is one of the first fruits (Figure 1-1). CCE is therefore based on the principles defined in CIE.

CIE was developed on the basis of the fundamental concept that cyber adversaries are well equipped and will be successful in attacking any system given enough time and resources. The evidence we continue to see of systems being compromised (including systems whose very design is to provide cybersecurity) has demonstrated that this concept is painfully accurate.

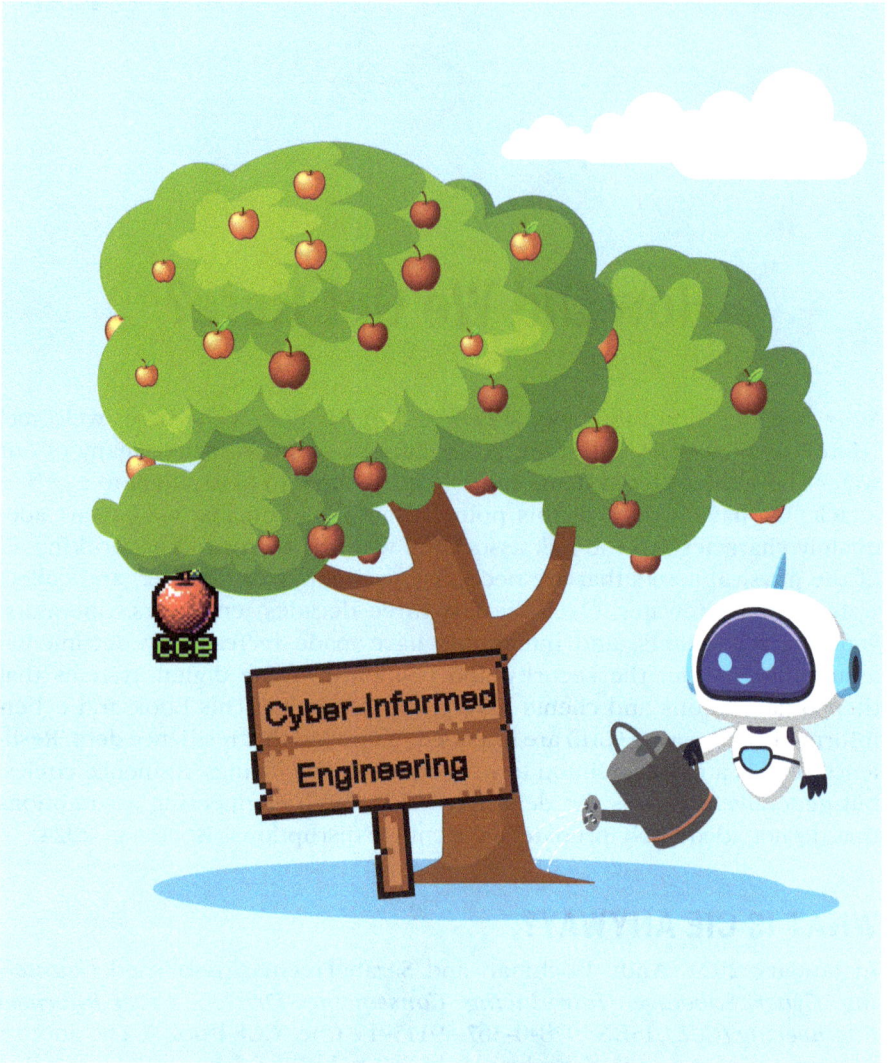

Figure 1-1   Consequence-driven, cyber-informed engineering (CCE) is one of the first fruits of cyber-informed engineering

CCE provides a detailed method to assess systems or design, especially relevant to complex systems such as those implemented by sovereign governments, including military operations.

CIE is therefore a design concept, and CCE is an implementation. This book focuses on CIE as the design approach and, at times, references methods used by CCE. Throughout this book, we will generally be discussing the design concepts of CIE unless otherwise stated. The CIE principles are explained in detail in the Cyber-informed Engineering Implementation Guide (INL 2023). Table 1-1 provides a summary of the 12 principles of CIE, adapted for the Water Sector.

Table 1-1   The 12 Principles of CIE

Principle	Definition
1—Consequence-focused design	Design water/wastewater systems to mitigate or minimize the consequences of cyberattacks on the utility's critical function/s. This incorporates design elements from disciplines including process, electrical, mechanical, and civil engineering.
2—Engineered controls	Identify and implement specific engineering design controls that reduce the need for "bolt-on" security later in the design and operations lifecycle phases.
3—Secure information architecture	Protect critical data and data streams and the processes they control through secure architecture practices such as network segmentation, data segregation, data replication, and role-based access.
4—Design simplification	A detailed explanation on how to "keep it simple" to reduce the consequences of cyberattacks while maximizing the benefit of technology for the utility.
5—Layered defenses	Establish an approach to protect critical functions through overlapping practices such as device/software diversity, redundancy, and system hardening.
6—Active defense	Design in both human elements (e.g., operators) and technology (e.g., OT network monitoring) to establish and maintain cybersecurity awareness as part of the overall system design.
7—Interdependency evaluation	Recognize and address the impact on critical functions of failures of systems outside of the design engineer's control. This includes, but is not limited to, electrical power and communications systems.
8—Digital asset awareness	Understand every digital device within the utility's system and how it supports or can be leveraged against the critical functions of the organization. This includes but is not limited to equipment and package systems provided by vendors, system integrators, and OEMs.
9—Cyber-secure supply chain controls	Use procurement language and contract requirements to ensure that vendors, systems integrators, and third-party contractors mitigate cyber-risk. Apply similar supply chain controls that the Water Sector has used for chemicals for many years. Apply equivalent controls for digital devices and related services.
10—Planned resilience	Assume your system will be compromised and design it to be successfully operated through the attack and recover safely.
11—Engineering information control	Prevent the authorized or unauthorized release of sensitive engineering records (e.g., requirements, specifications, designs, configurations, and testing) that could inform an attacker and improve their chances of carrying out a successful cyberattack.
12—Organizational culture	Build cybersecurity into the organizational culture to fully consider digital risks in system design and implementation.

PRIOR DISCUSSIONS OF CCE/CFA/CIE IN THE WATER SECTOR

The authors previously coauthored a *Journal AWWA* article titled "Engineering Cyber-Physical Resilience" (Ohrt et al. 2021). Highlighted in that article were three case studies on how utilities from around the country were embracing CIE concepts. Also highlighted in that article were the writings of Christian Lous Lange, who received the Nobel Peace Prize in 1921 (Figure 1-2). During his time, he advocated against new technologies that posed unique and increased risks to the international community if used to fight wars. A modern-day equivalent is the adoption of digital technologies within our critical infrastructure systems. Our engineering practices have evolved to depend on digital technologies, networking has become ubiquitous, and many features have been included in digital technology. This has resulted in numerous vulnerabilities and the potential for cyber-enabled failure modes, previously unaccounted for by engineering teams.

Source: Tønnesson et al. 2023

Figure 1-2 Christian Lous Lange

Evaluate the proper role of technology and dependence in a utility. Of course, this must be done relative to the critical functions of the utility and the expectations of the customers. It is unlikely that many utilities in the United States have the luxury of extended water service outages.

When this article was published in mid-2021, CIE was not close to the level of maturity that it currently is. In addition, relatively few utilities and organizations had begun reemphasizing the importance of manual operations capabilities.

REFERENCES

Bochman, A. 2024. Personal communication.

INL, 2023 Cyber-informed Engineering Implementation Guide. https://inldigitallibrary.inl.gov/sites/sti/sti/Sort_67122.pdf. Last Accessed: April 15, 2025.

Ohrt, A., D.A. Groves, J. Cox, A. Bochman, C. Cunningham, A. Hildick-Smith, and T.L. Kuczynski. 2021. "Engineering Cyber–Physical Resilience." *Journal AWWA*. 113(4):32-40. https://doi.org/10.1002/awwa.1708

Tønnesson, Ø., A. Sveen, and T. Greve. 2023. "Christian Lous Lange." Store norske leksikon. https://snl.no/Christian_Lous_Lange (accessed April 8, 2025).

Chapter 2

Why Is CIE Relevant for the One Water Sector?

THE CYBER-HYGIENE HAMSTER WHEEL

Cyber hygiene can be defined as "the practices and steps that users of computers and other devices take to maintain system health and improve online security. Much like personal hygiene, cyber hygiene involves regular, precautionary measures to protect systems from threats such as malware, phishing, and unauthorized access" (Brook 2023).

Examples of cyber-hygiene activities include

- regular software updates—updating operating systems, applications, and antivirus software;
- strong passwords—using complex passwords and changing them regularly;
- data backups—regularly backing up important data to prevent loss;
- network security and monitoring—using firewalls and secure Wi-Fi connections and specialized tools to monitor these systems for evidence of compromises; and
- user education—training users on recognizing and avoiding cyber threats.

Cyber hygiene is essential for all organizations. However, given the founding principle of cyber-informed engineering (CIE), these practices will not be enough to prevent a successful cyberattack because of the cyclical nature of the cyber-threat environment that could be likened to a hamster wheel (Figure 2-1).

This is not intended to make light of the substantial efforts that all sectors have taken around cyber hygiene. However, considering the assumption that all systems will eventually be compromised, CIE brings forward the potential to significantly reduce or eliminate the *consequences* of the inevitable cyber breach to our most critical water systems. Implementation of CIE during design can mean that a successful cyberattack against critical water infrastructure is a significant inconvenience instead of a disaster.

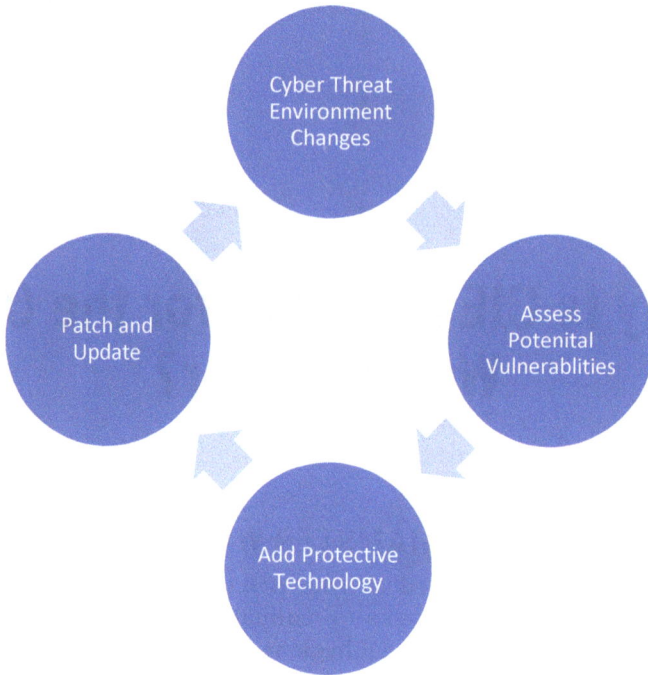

Figure 2-1   The cyclical nature of the cyber-threat environment

EVERYTHING INDUSTRIAL IS BEING DIGITIZED

In the industrial space, practically every type of equipment that engineers design into water systems has been digitized, in many cases with solid business drivers, including system visibility, efficiency, and cost. Equipment digitization is nothing new, and the pace of the trend is only moving forward more rapidly as technology advances. Table 2-1 provides a list of common equipment used in the water sector and how each example has been digitized.

Every digital device that we design into our critical infrastructure systems brings both opportunity and risk. Unfortunately, our engineering designs do not always account for the cyber risk these systems and components bring. For example, a key water treatment plant (WTP) may completely rely on an ultraviolet (UV) system as the primary source of disinfection, and the assumption is that the automated system can be relied on to manage the UV system to meet the disinfection requirements. *However, what would happen if the automation system were compromised in a way not obvious to the operations staff, simultaneously delivering water that has not been properly disinfected?* Does the design, operation, and maintenance approach address this type of failure scenario?

Table 2-1   Commonly digitized equipment used in the water sector

Function/Equipment	Digitized Examples
Motor controllers	Variable frequency drives Digital soft starters
Motor overload protection	Digital motor overload relays
Instrumentation	Wi-Fi–enabled pressure/flow transmitters IoT devices
Vendor systems (reverse osmosis, ultraviolet, chemical system)	Programmable logical controllers with 4G connectivity for maintenance/monitoring
Automated system	Programmable logic controllers
Electrical gear	Digital protection relays

THE INSIDER THREAT

An anecdotal story summarizes this threat well. The story goes that an individual whose job was to gain access to critical infrastructure systems around the world was developing his budget for the year. When planning for his budget based on the number of systems he was expected to compromise per year, he allocated just $15,000 per system. Why? That was the general going rate to get a disgruntled employee to provide unfettered access to *any* critical system around the world.

Although there are surely other anecdotal stories of loyal employees turning down much more lucrative offers to provide illegal access while doing the right thing and notifying authorities, the concept is clear. We must recognize that the compromise of key systems is inevitable, and we must design accordingly.

Cybersecurity for Critical Systems Is Not Just IT's Problem Anymore

Protecting the perimeter of an operations technology (OT) system alone is no longer viable because of the digitization of almost all key systems, the availability of communication options, and the well-known assumption of CIE. Therefore, the approach of maintaining a secure perimeter alone can no longer provide the protection needed for critical systems. We can no longer assume that our excellent cyber-hygiene practices will prevent a cyber breach. We need a different approach. That design approach is CIE.

REAL-WORLD INCIDENTS

Several reported cyber incidents in the water sector stand out when considering CIE as an approach to building cyber resilience. In each of these cases, the utility's automation was compromised. In most cases, the utility was able to adapt operations and ensure service to customers. Several of these are discussed in the following sections.

Aliquippa, Pa.—Nov. 25, 2023

The Municipal Water Authority of Aliquippa, Pa., experienced a politically motivated cyberattack by pro-Iranian hacktivists (Lyngaas and Sgueglia 2023). The Authority was using Israeli-made Unitronics OT equipment that was targeted by the hacktivists in the wake of the Israel-Palestine conflict that began in October 2023.

Operators responded to a loss of communications. After investigating and restarting the equipment, they observed the red screen with the hack notice. This screen was well publicized by victims of similar attacks from around the world and may be viewed here: https://bit.ly/Beaver-Countian. Utility staff manually operated the system, and there were no service interruptions to customers.

As reported by CNN, the Authority made the decision to replace the equipment as a precaution. Unplanned expenditures like this for a small utility can cause issues like deferred maintenance or delays on other improvements.

This incident demonstrates that in our connected age, water utilities are on the business end of geopolitics without any potential for recourse. Establishing capabilities, such as manual operations, that are inaccessible to cyberattackers is critical for utilities to maintain whenever possible.

Abernathy, Tex., and Muleshoe, Tex.—Jan. 18, 2024

On Apr. 17, 2024, *Wired* reported that the Cyber Army of Russia Reborn was actively targeting the critical infrastructure of the United States and allies opposed to Russia in the conflict in Ukraine (Greenberg 2024).

This hacktivist group is affiliated with Sandworm, one of the most notorious advanced persistent threats in the world. Sandworm's operations are prolific enough to have been detailed in the book *Sandworm: A New Era of Cyberwar and the Hunt for the Kremlin's Most Dangerous Hackers* by Andy Greenberg (2024) and more recently by Google Cloud (formerly Mandiant) (Roncone et al. 2024). Both of these resources provide extensive details on Sandworm's cyber operations in various countries.

It is possible that Sandworm is sharing capabilities with Russian hacktivist groups in an effort to spur critical infrastructure effects.

> "'Even though this group is operating under this persona that's tied to Sandworm, they do seem more reckless than any Russian operator we've ever seen targeting the United States,' [John] Hultquist says. 'They're actively manipulating operational technology systems in a way that's highly aggressive, probably disruptive, and dangerous.'" (Greenberg 2024)

The video screen capture of the attack reportedly taken by the attackers may be seen at the *Wired* article website (Greenberg 2024).

The attack on the Abernathy, Tex., water system resulted in water overflowing from a tank. This was reported by a local resident who observed finished drinking water spilling out of the top of the tank.

Luckily, service to customers was not affected. However, the attack on Abernathy was the first to result in physical consequences, although minor, for a water utility.

Oldsmar, Fla.—Feb. 5, 2021

It was reported that an attacker accessed the Oldsmar, Fla., WTP control system (CISA 2021a). The attacker increased the dosing of sodium hydroxide by several orders of magnitude. The on-duty operator noticed the increase and immediately took action to ensure there was no effect on water quality. Since then, there have been numerous reports about what may have transpired (Vasquez 2023). That discussion is not germane to this book. What is relevant is a CIE-related analysis associated with this incident.

First, this is an excellent example of having a trusted human in the loop. In our experience, utilities place a great deal of trust in their operations staff. This incident is a good demonstration of the value of having that trusted operator in the loop to ensure that the systems perform as desired.

Second, there has not been a formal discussion of the physical controls on the feeding of sodium hydroxide into the treatment train. This would take the form of an evaluation of the hydraulic capacity of the equipment to deliver a chemical solution. The physical infrastructure likely could not handle an increased chemical feed rate of that magnitude.

Third, this incident resulted in one of the most unique Cybersecurity & Infrastructure Security Agency (CISA) alerts ever issued (CISA 2021a). Titled "Compromise of U.S. Water Treatment Facility" and dated Feb. 12, 2021, immediately following the incident, it is the only CISA alert that we have seen to reference cyber–physical safety system controls, including the following:

- "Size of the chemical pump
- "Size of the chemical reservoir
- "Gearing on valves
- "Pressure switches, etc." (CISA 2021a)

This was a glimpse at a potential sea change in the conversation on how we manage risk to our water and wastewater systems. However, it was only echoed once, in an October 2021 CISA (2021b) alert titled "Ongoing Cyber Threats to U.S. Water and Wastewater Systems," which reiterated the considerations quoted earlier and added:

These types of controls benefit WWS Sector facilities—especially smaller facilities with limited cybersecurity capability—because they enable facility staff to assess systems from a worst-case scenario and determine protective solutions. Enabling cyber-physical safety systems allows operators to take physical steps to limit the damage, for example, by preventing cyber actors, who have gained control of a sodium hydroxide pump, from raising the pH to dangerous levels.

Unfortunately, additional conversations on water sector cybersecurity have emphasized cyber-hygiene approaches, which are important but do not address the engineering and operations-based approaches to reduce cyber risk.

Erris, Ireland—December 2023

Similar to the Aliquippa incident, the same Iranian hacktivist group accessed and cut off the water supply for 180 homeowners for nearly two days (Kovaks 2023, Quinn 2023). This utility was using a Eurotronics, an Israeli-made water pumping system. This was the first reported incident with a corresponding service disruption.

Cyberattacks—November 2023 to April 2024

To illustrate the escalation in the number of cyberattacks in the wake of the Russia-Ukraine war and the Israel-Hamas war, the Office of the Director of National Intelligence (ODNI) created the summary graphic shown in Figure 2-2. This includes 12 cyberattacks on water utilities around the United States. For each of the incidents shown on the graphic, including the

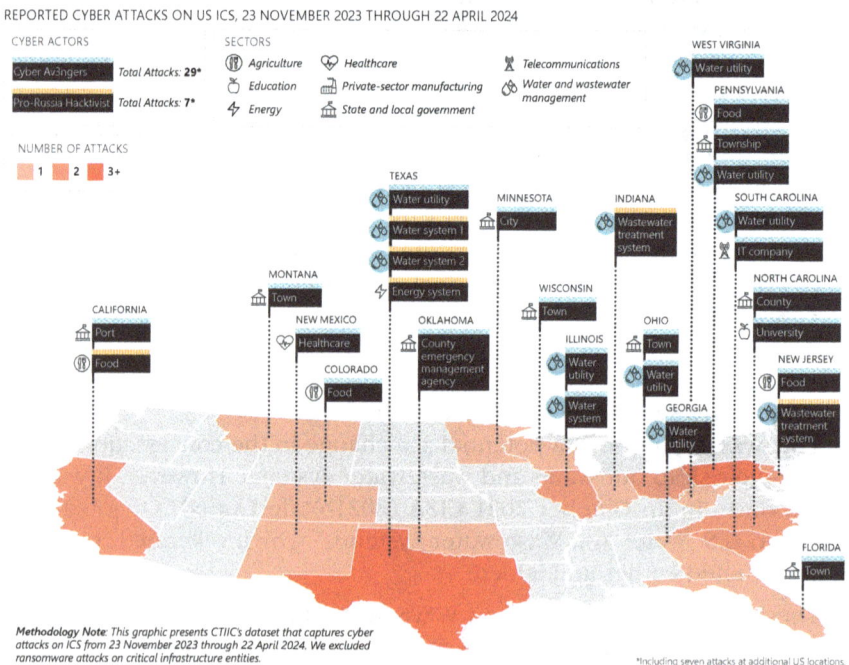

Source: ODNI CTIIC 2024

Figure 2-2   Recent cyberattacks on US infrastructure, November 2023 to April 2024
Note the 12 water sector cyberattacks during this six-month period. The water sector cyberattacks are denoted with a blue circle.

water sector cyberattacks, the "cyber actors gained access to and in some cases manipulated critical U.S. industrial control systems" (ODNI CTIIC 2024).

The summary states, "These attacks highlight a potential public safety threat and an avenue for malicious cyber actors to cause physical damage and deny critical services." This is an excellent summary of the cyber-threat environment in which water and wastewater utilities must operate. A primary concern is that the number of attacks and sophistication of these attacks will continue to increase as adversaries build a better understanding of utilities' cyber posture and operations. To support critical infrastructure owner/operators, the federal government provides advisories and alerts on a regular basis with insights in the nature of the attacks and how to prevent them.

ADVISORIES AND ALERTS

In response to the growing number of successful cyberattacks and the associated potential consequences of the attacks, federal agencies have issued numerous advisories in support of water and wastewater utilities and critical infrastructure in general. Many of these advisories are repetitive because the standard cyber-hygiene best practices are consistent. However, each one contains nuggets of helpful information that deserve acknowledgment and attention. These are presented chronologically.

"Control System Defense: Know the Opponent" (September 2022) (NSA and CISA 2022)

This is a uniquely excellent advisory, and portions of this are used by the authors in many conference presentations. One of the quotes highlighted within the advisory is that "traditional approaches to securing OT/ICS do not adequately address current threats."

In addition, the advisory highlights adversaries' planning process.

Cyber actors typically follow these steps to plan and execute compromises against critical infrastructure control systems:

1. Establish intended effect and select a target.
2. Collect intelligence about the target system.
3. Develop techniques and tools to navigate and manipulate the system.
4. Gain initial access to the system.
5. Execute techniques and tools to create the intended effect. (NSA and CISA 2022)

This is helpful because it allows for the articulation of controls and capabilities at each step. This can also be referred to as a "cyber kill chain." Within consequence-driven, cyber-informed engineering (CCE), analysts use the CCE kill chain, which is defined as a "visualization of the processes and steps an adversary takes for cyber-enabled sabotage."

The intended effects are described in the advisory in broad terms as the five D's of cyber sabotage—disrupt, deny, degrade, destroy, and deceive. As the authors have worked on various CIE projects; the 5-D construct is helpful in identifying the types of consequences of a cyberattack.

Consistent with CIE, the advisory states, "Understanding that being targeted is not an 'if' but a 'when' is essential context for making ICS security decisions." This helps frame up the certainty of cyberattacks and is helpful in CIE-related discussions.

"IRGC-Affiliated Cyber Actors Exploit PLCs in Multiple Sectors, Including U.S. Water and Wastewater Systems Facilities" (Dec. 1, 2023) (CISA et al. 2023)

This advisory summarizes the intrusions and associated mitigations to reduce the risk of the Iranian Government Islamic Revolutionary Guard Corps (IRGC)–affiliated group known as the CyberAv3ngers. The real-world incidents described from Aliquippa, Pa., and Erris, Ireland, are examples of the attacks carried out by this group. This advisory summarizes an excellent example of how geopolitical tensions can result in cyber incidents for water and wastewater utilities.

"[People's Republic of China] State-Sponsored Cyber Activity: Actions for Critical Infrastructure Leaders" (March 2024) (CISA 2024)

The advisory highlights the actions of a People's Republic of China state-sponsored group known as Volt Typhoon. The advisory warns "cybersecurity defenders that Volt Typhoon has been pre-positioning themselves on U.S. critical infrastructure organizations' networks to enable disruption or destruction of critical services in the event of increased geopolitical tensions and/or military conflict with the United States and its allies." This advisory highlights the strategic geopolitical nature of cyber intrusions. While an intrusion may not be for an immediate gain, if there is an international incident, critical infrastructure organizations may become targets in a relatively short time as compromised systems are manipulated to create the desired consequences for the US homeland.

WHAT IS CCE AND HOW IS IT RELEVANT?

CCE is the concept that actually introduced us to CIE. Andy Bochman (2018) wrote the provocatively titled piece "The End of Cybersecurity" in the *Harvard Business Review*. When Dan came across that, he sensed that this was the beginning of a change in how we view cybersecurity and resilience for critical infrastructure.

In the article, Andy rolls out—for the first time publicly—the CCE four-step methodology (Figure 2-3). CCE is a rigorous approach to identify and evaluate the worst "cyber-days" for any organization. CCE emphasizes focus

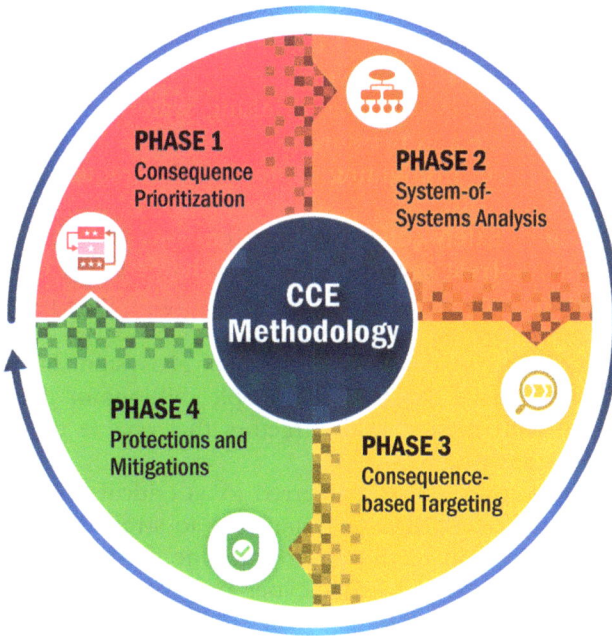

Source: Used with permission from BEA/INL

Figure 2-3   The four-step CCE process

on their "crown jewels" and acknowledges that cyber hygiene (e.g., strong passwords, network segmentation), while essential, is imperfect at best.

The assumptions for CIE noted earlier in this chapter hold for CCE as well:

- Systems will be subjected to successful cyberattacks.
- During an attack, systems will be unreliable and/or unavailable.
- All digital devices and systems are vulnerable.

One of the great things that has emerged from CCE is a set of concepts and definitions that have been helpful in communicating how utilities can take a different perspective on their OT cybersecurity practices. Some of our favorites include the following:

- Unverified trust—The blind reliance placed in people, processes, technology, information, and infrastructure (PPTII)
  - For example, a utility places trust in their integrator to securely configure network devices and use secure programming practices. However, in many cases, the utility doesn't have standards or the time and resources to look over the integrator's shoulder to ensure it should be done.

- Critical functions—The actions or activities that make up the organization's primary mission or purpose. For water and wastewater utilities, these are generally the following:
  - Water—Treat and distribute drinking water to customers at the required volume and pressure.
  - Water—Provide firefighting water at the required volume and pressure.
  - Wastewater—Safely collect wastewater.
  - Wastewater—Treat wastewater and discharge to the environment.
- Enabling functions—The PPTII the organization uses to both logically and physically deliver critical functions
  - Every utility is made up of a complex ecosystem of PPTII. Larger utilities will have hundreds of people, billions of dollars of infrastructure, and numerous contractors that enable the critical functions on a day-to-day basis.

More and more, CCE is being referred to as critical function assurance (CFA) (Gellner et al. 2023). This is being done to streamline the communication and drive adoption of the approach. One of the challenges is that when we use the term "cyber," non–cyber professionals take the mindset that cyber-related topics are "IT's problem" and aren't directly related to the mission and critical functions of the utility. Changing the terminology from CCE to CFA for many organizations is appropriate and helpful. Despite the name change, the key concepts, terminology, and approach are the same as those of CCE.

CASE STUDIES

Everything stated in this chapter is excellent background, but it is the authors' experience that the value of CIE can't be articulated without case studies. The following chapters and Appendix A provide numerous case studies. Each of these is based on real-world projects, assessments, exercises, and training that the authors have completed with our clients and colleagues.

Disclaimer #1

These case studies are fictionalized accounts of real water and wastewater utilities implementing elements of the CCE methodology. Names, locations, events, utilities, regions, countries, and incidents are fictitious. Any resemblance to actual utilities or events is purely coincidental.

Disclaimer #2

Any reference to specific equipment, vendors, or technologies in this study does not imply increased susceptibility to cyberattack over other brands or devices. The equipment noted is "typical" equipment often found in the industry. The principles presented in this work are universally applicable, regardless of the technology a specific utility might have.

REFERENCES

Bochman, A. 2018. *The End of Cybersecurity*. Brighton, Mass.: Harvard Business Review. https://store.hbr.org/product/the-end-of-cybersecurity/ BG1803 (accessed Nov. 27, 2024).

Brook, C. 2023. "What Is Cyber Hygiene? A Definition of Cyber Hygiene, Benefits, Best Practices, and More." *Fortra's Digital Guardian*, May 6. www.digitalguardian.com/blog/what-cyber-hygiene-definition-cyber-hygiene-benefits-best-practices-and-more (accessed April 15, 2025).

CISA (Cybersecurity & Infrastructure Security Agency). 2024. "PRC State-Sponsored Cyber Activity: Actions for Critical Infrastructure Leaders." Washington, D.C.: www.cisa.gov/sites/default/files/2024-03/Fact-Sheet-PRC-State-Sponsored-Cyber-Activity-Actions-for-Critical-Infrastructure-Leaders-508c_0.pdf (accessed Nov. 26, 2024).

CISA. 2021a. "Compromise of U.S. Water Treatment Facility." AA21-042A. Washington, D.C.: CISA. www.cisa.gov/news-events/cybersecurity-advisories/aa21-042a (accessed Nov. 26, 2024).

CISA. 2021b. "Ongoing Cyber Threats to U.S. Water and Wastewater Systems." AA21-287A. Washington, D.C.: CISA. www.cisa.gov/sites/default/files/publications/AA21-287A-Ongoing_Cyber_Threats_to_U.S._Water_and_Wastewater_Systems.pdf (accessed Nov. 26, 2024).

CISA, FBI (Federal Bureau of Investigation), NSA (National Security Agency), EPA (Environmental Protection Agency), and INCD (Israel National Cyber Directorate). 2023. "IRGC-Affiliated Cyber Actors Exploit PLCs in Multiple Sectors, Including U.S. Water and Wastewater Systems Facilities." AA23-335A. Washington, D.C.: CISA. www.cisa.gov/sites/default/files/2023-12/aa23-335a-irgc-affiliated-cyber-actors-exploit-plcs-in-multiple-sectors-1.pdf (accessed Nov. 26, 2024).

Gellner, J.R., C.P. St Michel, S. McBride, and M.R. Steffensen. 2023. *Critical Function Assurance: Understanding Critical Function and Critical Function Delivery is Foundational for Meaningful ICS Security Improvement and Policy Efforts*. INL/MIS-23-75497-Revision-0. Idaho Falls, Id.: Idaho National Laboratory. https://inldigitallibrary.inl.gov/sites/STI/STI/Sort_75387.pdf (accessed Nov. 27, 2024).

Greenberg, A. 2024. "Hackers Linked to Russia's Military Claim Credit for Sabotaging US Water Utilities." *Wired*, April 17. www.wired.com/story/cyber-army-of-russia-reborn-sandworm-us-cyberattacks/ (accessed Nov. 26, 2024).

Greenberg, A. 2019. *Sandworm: A New Era of Cyberwar and the Hunt for the Kremlin's Most Dangerous Hackers*. New York: Doubleday.

Kovaks, E. 2023. "Cyberattack on Irish Utility Cuts Off Water Supply for Two Days." *SecurityWeek*, Dec. 8. www.securityweek.com/cyberattack-on-irish-utility-cuts-off-water-supply-for-two-days/ (accessed Nov. 26, 2024).

Lyngaas, S. and K. Sgueglia. 2023. "Federal Officials Investigating After Pro-Iran Group Allegedly Hacked Water Authority in Pennsylvania." *CNN*, Nov. 28. www.cnn.com/2023/11/28/us/pennsylvania-water-cyberattack/index.html (accessed Nov. 26, 2024).

NSA and CISA. 2022. "Control System Defense: Know the Opponent." Washington, D.C.: NSA and CISA. https://media.defense.gov/2022/Sep/22/2003083007/-1/-1/0/CSA_ICS_Know_the_Opponent_.PDF (accessed Nov. 26, 2024).

ODNI CTIIC (Office of the Director of National Intelligence, Cyber Threat Intelligence Integration Center). 2024. "Recent Cyber Attacks on US Infrastructure Underscore Vulnerability of Critical US Systems, November 2023–April 2024." www.dni.gov/files/CTIIC/documents/products/Recent_Cyber_Attacks_on_US_Infrastructure_Underscore_Vulnerability_of_Critical_US_Systems-June2024.pdf (accessed Nov. 27, 2024).

Quinn, T. 2023. "Hackers Hit Erris Water in Stance over Israel." *Western People*, Dec. 7. https://www.westernpeople.ie/news/hackers-hit-erris-water-in-stance-over-israel_arid-4982.html (accessed Nov. 26, 2024).

Roncone, G., D. Black, J. Wolfram, T. McLellan, N. Simonian, R. Hall, A. Prokopenkov, D. Perez, L. Aytes, and A. Wahlstrom. 2024 *APT44: Unearthing Sandworm*. Alexandria, Va.: Mandiant. https://cloud.google.com/blog/topics/threat-intelligence/apt44-unearthing-sandworm (accessed Nov. 26, 2024).

Vasquez, C. 2023. "Did Someone Really Hack Into the Oldsmar, Florida, Water Treatment Plant? New Details Suggest Maybe Not." *Cyberscoop*, April 10. https://cyberscoop.com/water-oldsmar-incident-cyberattack/ (accessed Nov. 26, 2024).

Chapter 3

A Day Without SCADA Case Studies

CASE STUDY # 1: RECOVERING THE LOST ART OF LOCAL MANUAL OPERATIONS

Introduction

Imagine being a passenger on an airplane in which the flight crew has never manually flown or landed a plane in 20 years. Would you feel safe? One municipality in the western states recently found itself in a similar situation with its operational dependency on automation when it came to the function of backwashing filters, specifically, the dependency on the water treatment plant (WTP) supervisory control and data acquisition (SCADA) system conducting that task.

During a recent tabletop exercise focused on continuing operations in the absence of automation, the lead operator asked the question: Do we as operators know how to manually backwash the filters in the event of a cybersecurity breach that would disable our SCADA system for the WTP? The answer was overwhelmingly "no." Not one of the operators present knew what steps to take or had any documentation as to how to manually backwash the filters. They had never needed to know how to do this.

This dependence on automation wasn't something that happened overnight; in fact, it was a long dependency on automation in which generations of operators from the plant had retired and manual operation of the plant had ceased, even before the youngest operator present was born. Backwashing was part of the operator's duty, but it was a matter of pushing a button and monitoring the SCADA screen outputs to ensure the backwash process was going according to the program's instructions.

As WTP capital improvements were being conducted over the years, design engineers phased out the manual backwash system capabilities for more efficient and optimized automated backwash systems. This included the removal of old backwash pedestals in which operators would manually control backwash functions and change out the valves, actuators, and control switches in places that were relatively unknown to operators. The timing of all of the backwash sequences worked on their own, and relatively well, so there was no need to manually operate the systems. Routine maintenance was conducted on critical parts, which kept the operator's curiosity and concerns "out of sight, out of mind."

Planning

The operations team decided it needed to tackle the issue of manually backwashing filters and asked the first question: Was it even possible to do so given the configuration of the treatment plant and location of key equipment and instruments to conduct a backwashing sequence? Certainly, one person could not do it alone, but they did not know even how many operators they needed to conduct a backwash cycle.

The first step the team took was to identify and deconstruct the backwash programming data in the SCADA system. This information would provide what timing of valve opening, throttling, and closing. Target flow rates and target pressure ranges are needed to effectively and safely backwash a filter. The team then surveyed the locations of each valve within the plant that were involved in the sequencing, points of monitoring needed, and access requirements because some of these valves and other points of local control were not immediately accessible. To access one critical analog gauge, the team had to purchase ladders to get staff in the right position for monitoring.

Developing a Playbook

General sequencing, transitions, and communications were tested by a group of operators on a dry run. In all, six operations staff were needed to develop and test out the manual backwash operations at first, but the lead operator for the plant was confident the number could be reduced to four staff after some practice. A total of 15 sequential steps were identified as required to safely backwash a filter and place it back into service. The other goal was to keep the overall time to conduct the backwash in 30 min, the same as the automated process.

Manual Backwash Exercise

Given the team wanted to prove an actual manual backwash, they set some practical parameters to conduct their exercise:
- Do it at a time of year when other system demands are minimal. They didn't want to disrupt operations or service to customers in any way.
- Schedule and conduct the exercise during normal business hours.
- Don't stress staff. This was recognized as a learning and training process.

They also had a senior operator act as a "quarterback" to call out the step signals and at a place where they could visually confirm process changes within the filter gallery. Each of the six staff members used radios to communicate each step to each other to ensure continuity of sequencing. The planned exercise was a success with many lessons learned as they conducted their hot wash meeting of the exercise.

Lessons learned: Operations

- Manual backwash was possible but needed to be coordinated, and at least four people were estimated to be needed to conduct the sequencing. There are only two staff on shift at night, so additional staff would need to be called in if needed.
- All senior staff should know how to operate a plant manually. These skills should be passed down through training and mentorship to junior operators.
- Additional analog gauges and more accessible local control status monitoring points are needed to monitor status.

Summary

Automation of any process can increase efficiency and provide opportunities to optimize processes and should be used whenever possible. However, automation versus manual operation should never be an either/or proposition. Management will implement automation as a labor-savings tool, and operations will use it for optimization to their benefit. For this reason, the return on investment is usually in favor of implementing and using automation.

There is, however, an emerging realization that the water industry's dependency on automation tools has a fatal flaw: We trust in it so much that we are losing key operations knowledge and the engineering practice of providing redundant and resilient system designs for critical assets. Also, with the increase in disruptive cybersecurity malevolent acts, power outages, and communications issues, the reliability of automated systems is now suspect, and utilities must be ready to operate in manual mode on the fly.

Unwinding dependence on automation can be achieved! Manual operations and automation are complementary to each other, and neither system is 100% reliable on its own, but together, complete optimization and resilience of critical system processes can be obtained.

CASE STUDY #2: LOOK MA, NO HOAs!

An advanced water sector engineering firm based in the western US was contracted by a regional water authority to serve as the owner/advisor for a project to deliver a resilient, sustainable water supply supporting the integrated use of groundwater and surface waters to diversify the region's water supply portfolios. Water is pumped from a river via an infiltration gallery, through an intake structure, source water pump station, and pipeline to be treated at a new treatment plant and delivered to nearby cities. This complex project required the development of a strategy for the planning, design, construction, and operations of these new regional surface water facilities, and local distribution infrastructure necessary to connect with each city's existing system. The western US engineering firm provided program management and served as owner/advisor for planning, technical studies, preliminary design, program development, and procurement document preparation for the design-build delivery model used to design and construct the new regional facilities. This project delivers 15 mil gal per day (mgd) of drinking water, securing a sustainable and resilient supply for customers.

Introduction: How Did We Get in the Jam?

During the design of the project, the engineering team included the following in the Basis of Design Report: "Motor Control Center (MCC) overloads shall be electronic type with communications networking capability. . . All networkable components shall be connected to a central network switch to facilitate communications with plant control system."

"MCC overloads" refers to the equipment that monitors electrical power being supplied to equipment including motors and disconnects the load to prevent electrical overload of equipment. In addition to providing motor overload control, the overload selected (Allen Bradley E300) also provides digital output contacts that could be used for multiple purposes, including signals to start and stop the motor in manual mode.

The Problem: What is the Jam?

However, when the design was reviewed from a cyber-informed engineering (CIE) perspective, the question was asked: How will this facility be run in the face of a successful cybersecurity attack via the communications network that made any or all of the E300 relays unavailable or unreliable? The answer from the engineers was that it would be impossible to run any motors in the facility, even in a "manual" mode. Effectively, this meant that there were no "hand-off-auto (HOA)" switches available to operations in the case of a system malfunction and would require a complete rewire of the MCC to operate any equipment (Figure 3-1).

The Solution: How Did We Get Out of the Jam Using CIE Principles?

Considering the following applicable CIE principles helps engineers:

Consequence-focused design: First, by recognizing that cyberattacks will occur, we now consider a cyberattack as an additional failure mode that we need to design to address, instead of assuming that a cyberattack would be impossible or unlikely.

Digital asset awareness: Understanding the proposed role of the overload relays and how they were being used allowed the engineers to recognize the effect of a cyberattack failure of these devices and the effect on the facility, which would result in a high-consequence event that rendered the facility inoperable.

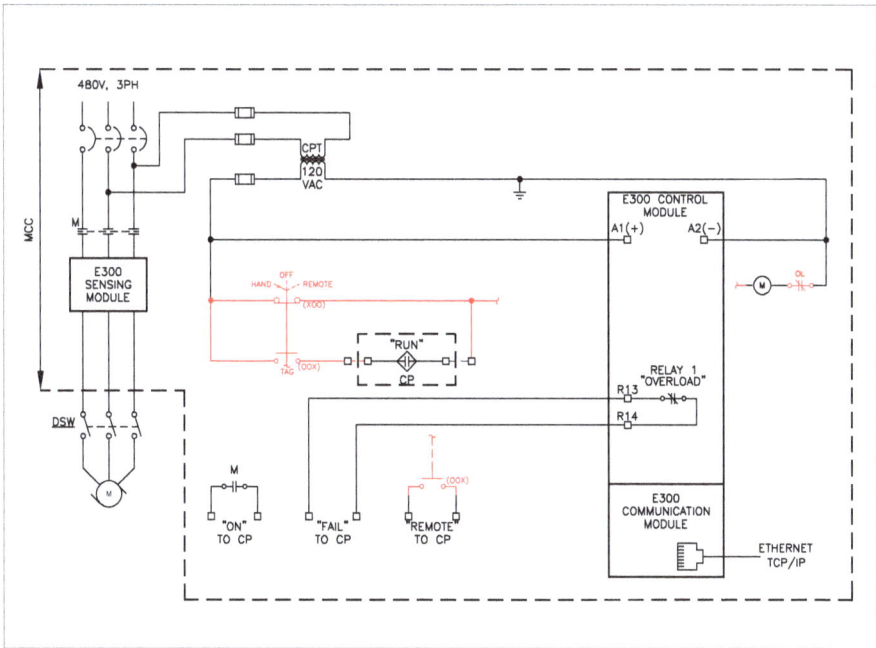

Figure 3-1 MCC wiring diagram with HOA control addition

Planned resilience: By planning for the cyberattack scenario, the engineer and operations team recognize the cyberattack failure mode and are prepared to operate the facility using the included (HOA) switch in the case of a failure of the overloads. Of course, in order to do this, qualified electrical staff will need to understand how to bypass the overload if required during an emergency situation that prioritizes system operation over potential damage to an individual equipment component. Fortunately, the electrical bypass is a simple wiring change, instead of a complete rewire of the motor starter.

CASE STUDY #3: WASTEWATER TREATMENT PLANT IN NORTHERN CALIFORNIA—WIRING OF INFLUENT PUMP MCCs

A wastewater treatment plant (also called the water pollution control plant) located near a local landfill is owned and operated by a city in Northern California. The facility is permitted to treat 7.5 mgd of wastewater. The existing treatment system design capacity is 6 mgd based on average dry weather flow. There are two permitted discharge points from the plant. Treated effluent is discharged to the local overflow channel and drain.

One key component of the plant is the influent lift station, which lifts the wastewater from over 70 ft below grade to the headworks portion of the plant at grade. The structure where these pumps are installed includes a large vertical structure with its top at grade level and a grate allowing operators to see the incoming wastewater as well as the large pumps responsible for lifting water into the plant. If the pumps do not lift the wastewater, water could begin to fill up the structure, and raw wastewater would soon flow to ground-level diversion ponds adjacent to the plant. Depending on the flow into the plant, there is the potential that the influent pump station structure could also be flooded in the process.

At the top of the structure are several (one for each pump) HOA switches that allow operators to manually start and stop the pumps. These are intended to be used only when manual operation of the pumps is required, which is usually only during maintenance operations.

Introduction: How Did We Get in the Jam?

When the electrical engineer was designing the system, one challenge that was encountered was the distance from the HOA switches to the electrical gear that controlled the pumps, which was located two floors below ground level. That distance created a need for significant amounts of conduit and wire between the HOA switches and the MCC. Another issue considered by the electrical engineer was the cost of the MCC buckets, which provided both control equipment and control power transformers for electrical control circuits used by the HOAs.

The combination of these challenges led the electrical engineer to make two key changes to the design. First, the engineer decided to remove all control power transformers from the MCCs and provide control power from the programmable logic circuit (PLC) panel, which was collocated with the MCCs. This approach was intended to save costs for the MCC buckets. Second, the engineer decided to have the HOAs provide input to the PLC, which in turn would provide an output to the MCC buckets. This approach saved a significant amount of conduit and cable over the traditional approach of wiring the HOAs directly to the MCC buckets.

The Problem: What is the Jam?

While the approach chosen for this design likely provided some cost savings, it introduced a series of potential consequences not foreseen at the time. The consequences can be summarized by asking the following question: What happens if the PLC is either unavailable or unreliable?

When this question was asked of operations staff, their immediate reaction (without knowing the details of the electrical design) was: "We just put things in manual and operate the pumps from the top of the influent station." To put this approach to the test, during a plant outage, a test was developed based on a cyberattack that caused the pumps to stop operating.

It was decided to power down the PLC, and operations staff located themselves at the top of the influent structure and at lower levels of the station where they could manually monitor levels. The plan was to communicate via radio the need to start and stop the pumps manually while the PLC system was repaired and/or replaced.

Guess what happened when the PLC was powered down and the operators tried to start the pumps? The pumps did not start. Given the mechanical structure and process, this scenario, if it occurred during a high-flow time, could be disastrous for the plant.

The Solution: How Did We Get Out of the Jam Using CIE Principles?

In this case, there are several CIE principles that would have helped during the design to avoid the situation that will also help after the fact to design our way out of the current dilemma. These changes involved installing control power transforms in each of the MCC buckets and wiring the HOA switches on the top deck of the lift station to be electrically tied directly to the MCC buckets that controlled the motors. This meant that as long as electrical power was available to the MCC, the motors could be controlled manually from the HOA switch. No PLC power or control is required for manual operation with this design change.

Applicable CIE Principles

Consequence-focused design: The designer needed to recognize what the potential consequences of a failure of a malicious attack against the PLC would mean for plant operations.

Engineered controls: Implement basic controls that allow the system to continue to be operated manually without the digital devices that are subject to sabotage. One potential improvement would be to provide float switches and basic local controls that would start and stop pumps based on the water level in the lift station even in the absence of the PLC system.

Design simplification: Although the original design approach lowered some construction costs, it increased the complexity of the system, which provided an avenue that an attacker could leverage.

Interdependency evaluation: It is critical for engineers and operations to understand the interdependencies between systems. Removing the control power transformers from the MCC buckets saved some construction costs but created an electrical interdependency with the PLC panel.

Digital asset awareness: The recognition that all digital assets, including PLCs, are vulnerable to misuse would lead the engineer away from relying on the PLC for manual control of the pumps.

Planned resilience: Planning for system failures would likely cause the engineer to provide a backup control system using simple float as a backup to PLC control during a system attack or outage. This approach would save valuable time for operations while the PLC system was being restored to reliable operation.

Chapter 4

CCE/CFA Case Studies

CASE STUDY #4: DOING CIE/CCE WITHOUT REALIZING IT

Introduction

ACME Water operates nine unconnected public water systems covering nearly 100% of the population of an island. The system has 50 groundwater wells and over 500 miles of distribution pipeline. Water from each groundwater well is chlorinated, and the water from several wells also requires treatment with soda ash. The distribution system includes 40 tanks that maintain pressure in the distribution system. The system pumps, treats, stores, and delivers approximately 15 mil gal per day of drinking water to customers.

The distribution system covers a largely rural area, including areas with limited communications coverage (3G/4G/5G). The hilly terrain presents challenges for private, line-of-sight radio communications. The distribution system is pressurized by the tanks situated at various elevations to provide sufficient system pressure via gravity. Each of the nine public water systems is controlled centrally from an ACME Water facility.

In addition to the challenges with establishing reliable communications, being on an island in the ocean, ACME Water's supply chains are long and sometimes unreliable. Therefore, staff were faced with a decision on how they could meet the reliability needs of their customers and achieve the desired level of convenience for their operations and maintenance staff. Serendipitously, these measures also resulted in resilience to cyber-physical incidents.

Personnel

In 2011, the current supervisory control and data acquisition (SCADA) manager joined ACME Water after serving as a US Navy submariner. The SCADA manager's experience in the Navy dealing with nuclear-powered submarines taught him the importance of having redundancy of controls. The SCADA manager reviewed the existing system and implemented a design approach that provided an excellent, reliable backup control mechanism that protected ACME Water from a potential communication outage.

The Problem

Before the SCADA manager joined ACME Water, the SCADA system was over 10 years old and not reliable. This resulted in a strong culture of manual operations of the distribution system because operations staff were required to resort to this method of operation. Manual operation capabilities were and still are an important part of the overall operations culture at ACME Water. In addition, because of ongoing operational challenges from the conditions noted previously, ACME Water had to take a nontraditional approach to threading the needle between reliability and convenience.

The Solution

To do this, ACME Water is engineered in out-of-band, nondigital controls to reliably control the booster pumps that lift water to the elevated reservoirs based on historical consumption patterns and system hydraulic limitations. This system is normally not used but can be activated with a "flip of a switch" at each of the booster stations and deep-well pumps. When active, this system keeps water flowing to the distribution system with no communications whatsoever required.

When implementing this backup operational strategy, ACME Water assessed the consequences of a less-optimized operational strategy. Recognizing that an improperly set timer may result in underfilling or overfilling of a reservoir, ACME Water opted to allow for potential overfilling of a reservoir. This risk to the staff, facilities, and customers is minimal because of physical security controls at the facilities.

ACME Water has been required to use this approach to operate significant portions of the system using this control approach because of several reasons. Although the system is not optimized in the same way a modern communications system is, it allows ACME Water to continue to meet its mission in the face of a communications outage and would serve the same purpose during a successful cyberattack on the SCADA system that renders the SCADA system unreliable or unavailable.

Applicable CIE Principles

This approach implements key elements identified by cyber-informed engineering (CIE) as indicated in *italics*:
   1. *Consequence-focused design*
   2. *Engineered controls*
   3. *Secure information architecture*
   4. *Design simplification*
   5. Resilient layered defenses
   6. Active defense

7. *Interdependency evaluation*
8. Digital asset awareness
9. Cyber-secure supply chain controls
10. *Planned resilience with no assumed security*
11. Engineering information control
12. *Security culture*

ACME Water intuitively developed a solution despite the lack of a formal CIE framework. We anticipate that ACME Water will continue to leverage CIE as a method to reinforce the importance of the design approach taken and continue to make sure that as the ACME Water system is upgraded over time, the design principles are maintained.

CASE STUDY #5: VIRGINIA WATER UTILITY—PROCESS INTERLOCKS

Introduction

An advanced water sector engineering firm based in California and serving other western US states had the opportunity to conduct a CIE review for a wastewater plant during its 90% design. This case study will present the methodology used to conduct the review, the findings, and the CIE recommendations given. This case study will focus on process controls and operational equipment technology. This case study will not focus on networking and/or traditional cyber-hygiene practices related to information technology equipment.

It is important to note this was a late stage in the design process to do a CIE review. Also, not all design documents were provided, such as process and instrument diagrams (P&IDs). Because of these circumstances, the CIE review was limited to what was provided.

CIE

Recently, the risks posed to critical infrastructure organizations through malicious attacks on control system technologies have become better understood. New approaches to managing this risk are continuously being developed, with CIE being the most comprehensive and rigorous. CIE is an emerging method to "engineer out" cyber risk across the engineering life cycle.

Based on the widespread vulnerabilities in our digital environments, there are three key assumptions underlying CIE:
- Systems will be subjected to successful cyberattacks.
- During an attack, systems will be unreliable and/or unavailable.
- All digital devices and systems are vulnerable.

To address these, CIE promotes the use of design decisions and engineering controls to mitigate or ideally eliminate the potential for a cyber-enabled failure mode to occur within an engineered critical infrastructure

Table 4-1   The 12 CIE principles and groupings for the water sector

Design and Operations	Organizational
Consequence-focused design	Interdependency evaluation
Engineered controls	Digital asset awareness
Secure information architecture	Cyber-secure supply chain controls
Design simplification	Planned resilience with no assumed security
Resilient layered defenses	Engineering information control
Active defense	Cybersecurity culture

system—reducing reliance on after-the-fact cybersecurity controls. Integrating cyber resilience into early-stage engineering decisions often delivers lower-cost, more sustainable, and more effective cybersecurity approaches. The early stages of engineering design are also when seemingly innocuous decisions can inadvertently permit the potential for cyber-enabled failure modes to persist within the system. This results in unknown and unmitigated risks that operators and engineers must monitor for and defend against in the future.

In August 2023, the US Department of Energy released the *Cyber-Informed Engineering Implementation Guide*, which is the first document to articulate what an implementation of CIE entails (Wright et al. 2023). The document presents the 12 CIE principles, shown in Table 4-1. While not official groupings, the 12 principles are split into two groups, Design and Operations and Organizational. These groupings are beneficial to the Water Sector to better communicate how the CIE principles relate to each other and in which phases of the engineering lifecycle they are most useful.

The CIE implementation guide provides an extensive articulation of these principles along with the water/wastewater case study included in Appendix C. This reference provides the most comprehensive approach to CIE to date. It is expected that the industry's understanding and approach to applying CIE will evolve quickly in the next five to 10 years. The industry and the western US engineering firm are just beginning to apply CIE within engineering design projects. The approach and subsequent results are reflective of our current and best application of CIE within the engineering design process.

Methodology

For this case study, the western US engineering firm was provided with control descriptions of the various wastewater processes in the plant. In each control description, hardwired and software interlocks were identified. The review evaluated each of these interlocks. Interlocks fall under the CIE principle of "engineered controls." These controls are often used to protect the equipment and the process. Network sheets were provided that identified variable-frequency drive (VFD) and reverse voltage soft starter (RVSS) connections. To confirm how VFD and RVSS were wired for control, wiring schematics were also reviewed. Lastly, to ensure there were no access points the utility didn't control, such as vendor remote access, vendor control panel electrical drawings were reviewed.

The western US engineering firm's ability was affected by not having the P&IDs and interviews with operators about the effects of design decisions and potential recommendations on each process. Assumptions had to be made based on the information available.

Lessons learned that would improve the CIE review:
- Conduct CIE review earlier in the design stage.
- Obtain and review P&IDs.
- Conduct workshops with operators to identify critical processes to the plant and how design decisions and recommendations would affect their ability to operate the processes.

Findings and Recommendations

Hardwired interlocks. This design did well at implementing hardwired controls that are intended to protect motors and equipment from accidental and intentional misuse.

Examples of hardwired interlocks include the following:
- Hardwired logic in the motor control center (MCC) or VFD shall shut down a running clarifier rake in response to any of the following conditions:
  - Torque high
- Hardwired logic in the VFD shall shut down a running primary sludge pump in response to any of the following conditions:
  - Discharge pressure high
  - Motor temperature high
- Hardwired logic in the VFD shall shut down a running recycle pump in response to any of the following conditions:
  - Suction pressure low
  - Motor temperature high
- The equalization basin mixers will shut down by hardwired interlock in response to any of the following conditions:
  - Motor temperature high
  - Moisture level high

As can be seen in the various types of equipment, hardwired interlocks are in place to protect the equipment. There were no recommendations given related to hardwired interlocks.

Software interlocks. There were software interlocks identified in the design that failed CIE and "engineered controls" that could give a malicious actor control over manual operations.

Examples of software interlocks include the following:
- At least one grit removal unit must be in service at all times. The programmable logic circuit (PLC) will not allow an operator to close gates in MANUAL control mode such that no units will be in service.

- At least one screening channel must be in service at all times. The PLC will not allow an operator to close gates in MANUAL control mode such that no channels will be in service.
- If either the influent or effluent gate within a screening channel is closed, the PLC shall not permit the operation of any component of the associated fine screen in MANUAL mode.

These software interlocks give the PLC the ability to override manual control. This would give a malicious cyber actor the ability to stop operators from being able to control the plant manually. It's assumed the designers wanted to protect the system from operator error, but this can limit the operator's ability to run the plant and cause consequences. When in manual, let the licensed operator have control of the system. It was recommended to move these software interlocks and to never override manual control with the PLC.

Lessons learned that would improve the CIE review:

- Have a discussion with operators about how proposed software interlocks affect the system and their ability to operate.

Network sheets, VFD, and RVSS wiring schematics. The network sheets did not provide any network configuration information; only the network architecture and all VFD and RVSS soft starters were connected via Ethernet. This led to a review of VFD and RVSS wiring schematics for one process.

What was found is the design used Ethernet for diagnostics and control of the VFD and or RVSS. When a signal from the field is taken directly to a solid-state, programmed device and used to activate an alarm that would be used to interlock the device, a malicious actor may change the set point or parameter in the solid-state device that could change the intended function. This is especially important when VFD and RVSS are connected to an Ethernet network.

The recommendation provided was to only use Ethernet for diagnostics and hardwired input/output (I/O) for control. Having hardwired I/O results in the ability to isolate and continue operating critical processes in the event of a cyberattack.

Lessons learned that would improve the CIE review:

- Conduct the CIE review earlier in the design stage.
- Obtain and review P&IDs.
- To properly review the network for CIE, conduct interviews with the network administrators.

Vendor control panel electrical drawing. In the review of the provided vendor control panel electrical drawings, no remote communications that would give the vendor remote access to the equipment were identified. This is important because it stops connections the utility might not be aware of, which can directly give a malicious actor control of a process. Although there were no remote communications found in the drawings, this does not mean a vendor won't add it or is unable to. To prevent this, it was recommended that the utility review its contractual documents and update them if need be.

Lessons learned that would improve the CIE review:
- Conduct the CIE review earlier in the design stage.
- Review SCADA standards and contractual agreements.

Summary

The CIE review was successful in identifying potential cyber-enabled consequences to the system. The major takeaway from this review is that although it's a 90% design, there is still time to make engineering changes to the design that will add cyber resilience to the process. Ideally, CIE is implemented from the very beginning of design. Make a design standards document at the beginning of the design that uses CIE principles, such as Ethernet for diagnostics, and hardwired I/O for control. Another is to never override manual control of the process with the PLC. The earlier these design standards are set in stone, the less financial cost there will be to make CIE changes later in design. A great way to implement CIE that will save time and affect design changes is to have all designers and engineers read the CIE implementation guide before beginning any new project.

CASE STUDY #6: SOUTHERN CALIFORNIA WATER DEPARTMENT—CCE IN SODIUM HYPOCHLORITE OPERATIONAL STRATEGIES

Background

The water department of a Southern Californian city operates 13 well pumps, each equipped with a sodium hypochlorite (NaOCl) dosing system. The sodium hypochlorite dosing system is used to treat water by maintaining the required chlorine residual before sending water either to a storage reservoir or directly into the distribution system. Currently, the sodium hypochlorite system is not integrated with the SCADA system and uses hardwired local controls to maintain dosing. When the well pump's hand-off-auto (HOA) selector switch is set to auto mode and the well at the site is started, the sodium hypochlorite pump is issued a command to start. This is accomplished through an interposing relay in the well MCC. The interposing relay activates the sodium hypochlorite pump through hardwired control when the well pump is called to run, which eliminates a vulnerability if an attacker were to gain control of the PLC and issue a start command instead. See Figure 4-1. When the pump is in hand mode, it is automatically called to run. The operator enters a sodium hypochlorite set point at the sodium hypochlorite metering pump. The sodium hypochlorite metering pump feeds the appropriate amount of sodium hypochlorite to meet the sodium

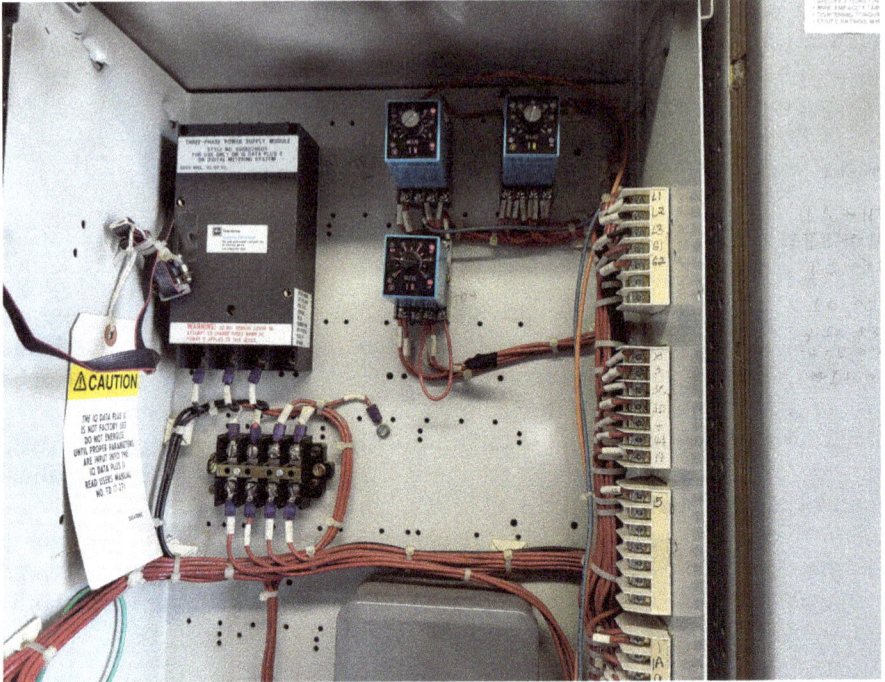

Figure 4-1 Interposing relay

hypochlorite concentration target. The city is in the process of upgrading its SCADA system and integrating the sodium hypochlorite dosing system into its SCADA system. This will allow the system to flow pace for control of the sodium hypochlorite pumps. With flow pacing, the PLC monitors the well discharge flow and adjusts the sodium hypochlorite pump speed based on the output from the well pump. This results in the speed of the pump being controlled proportionally based on the well flow rate to the reservoir and very accurate dosing.

High-Consequence Event

If the PLC is given full control over the sodium hypochlorite dosing system, and an adversary gains access to the SCADA system and targets the dosing system, there could be malicious modifications to the dosage control. An adversary could modify the PLC logic or dosing set points so that either no sodium hypochlorite dosing occurs or the dosing could be increased to a level that may result in unsafe conditions for customers and staff. The attacker could also ensure that no human–machine interface indications or alarms are triggered when the required residual is not met or when there are speed or flow discrepancies. As a result of the attack, either untreated or overtreated water could be distributed to and consumed by customers. Improperly treated water can cause serious, widespread illness and will require a system shutdown and extensive flushing operations. Critical customer service may be affected for several days. This event may lead to substantial financial and reputational damage to the utility.

Mitigations and Protections

To provide protection to the sodium hypochlorite system from a potential cyberattack, each well pump will be equipped with redundant metering pumps and dual receptacles to allow both pumps to be plugged in for rotational control. A three-position selector switch at the MCC will provide the system with the ability to run pump 1, alternate the pumps, or run pump 2 (P1-Alternating-P2). This functionality will be handled locally through an alternating relay, which would alternate the operation of the two pump loads. The PLC will have minimal control of the sodium hypochlorite system, meaning it will only control flow pace modes and receive feedback from the system. Ultimately, the sodium hypochlorite system can function without the PLC.

In the event of a PLC failure or an attack on the PLC, the well pump can run in hand mode, and the appropriate sodium hypochlorite metering pump will still start. In this scenario, the city would lose the ability to flow pace the sodium hypochlorite metering pump from the PLC. When the PLC is down, a configured alarm would dial out to call the operators and trigger local control of the metering pump's speed from the keypad at the pump skid. See Figure 4-2.

In the event of a PLC failure or an attack on the PLC, water will still be properly treated.

If the start command originated from the PLC, operators would have to remember to manually activate the pump and set sodium hypochlorite dosing. Once it was established that the well starter would issue a run command to the sodium hypochlorite pump, the next logical question was, "Which of the two sodium hypochlorite pumps should be started?" The project team considered simply wiring pump 1 or pump 2 to the interposing relay in the MCC. The issue with this approach is that if the pump has failed, the interposing relay would not start any sodium hypochlorite pump. Thus, the project team determined the start command and the pump rotation had to be handled through an alternating relay in the MCC. With this approach, if one of the sodium hypochlorite pumps has failed, the relay will issue a start command to the standby pump.

With the loss of the PLC, the sodium hypochlorite pumps will still be called to run and will be rotated by the well starter. Granted, the city will lose the ability to flow pace from the PLC. However, because the alarm dialer will notify operators of a PLC failure, the operators will maintain control over the dosing system by setting the dosing set point locally. Thus, they don't run the risk of distributing untreated or overtreated water.

Figure 4-2   Sodium hypochlorite metering pump keypad

CASE STUDY #7: SOUTHERN CALIFORNIA WATER DEPARTMENT—PROTECTION OF ROTATING EQUIPMENT

Protection of Rotating Equipment

When well pumps operate and pump water out of the well for treatment, the pump impeller and shaft spin in a certain direction. When a well pump shuts down, the water drains out of the riser (the pipe conveying the water up the well from the pump to the treatment works) and back down into the well. While this is happening, the impeller and shaft spin in the opposite direction. If an adversary turns the well pump back on and it spins in the original direction while water in the riser is placing pressure on the pump, the action could result in shearing the impeller off of the shaft because of the forces on it. This would result in extensive repair and placement costs, increasing with the depth of the well. In addition, that well would be out of service until repairs could be made.

When analyzed in the context of CIE, it is clear that a well-informed and determined adversary could easily manipulate set points or modify the PLC program to cause catastrophic damage to the well pump. This could be accomplished by starting and stopping the pump in rapid succession or changing direction on the motor controller.

To prevent catastrophic damage to their rotating equipment, the city has taken the following measures:

- Ratchet plates are installed on each of their well pumps. The ratchet plate is a physical, mechanical appurtenance attached to the well pump that prevents the pump from spinning backward. Even the most determined cyber adversary would not be able to overcome the ratchet plate remotely. Even if an adversary were able to manipulate set points or modify the PLC program, the ratchet plate would prevent this operation from occurring.

- Backspin timers are installed in the MCC buckets. An adversary can cause significant damage to rotating equipment by issuing stop commands followed by start commands in rapid succession. To prevent this scenario from occurring, it is common practice to use a backspin timer. The timer is initiated once a stop command is issued. The pump is interlocked until the backspin timer has elapsed. Unfortunately, the backspin timer is usually integrated via logic in the PLC. A determined and well-trained adversary can easily modify the PLC program and remove the backspin timer.

   The city has installed the backspin timer as a physical time-delay relay installed in the MCC bucket. With the backspin timer being a physical device hardwired into the motor start–stop circuitry, any modifications to the PLC program won't be able to bypass the hardwired backspin timer.

- Manual functionality is integrated at the MCCs and VFDs. Each well pump and booster pump has all the controls in the field to accomplish manual control, including start, stop, and sequencing. Each pump's MCC is equipped with an HOA switch wired independently of the PLC. This functionality allows operations staff to run equipment manually without the PLC. Critical interlocks are hardwired into the pump's start–stop circuitry.

  Critical feedback devices, including discharge pressure transmitters and flowmeters, are strategically located in the field so that operations staff can see them when operating equipment in manual. In addition, each field instrument is equipped with a local display so that the process variable can be seen.

Pumping Down the Water Table

The city's water distribution system is separated into two operational zones—East Zone and West Zone. The elevation is significantly higher in the northeast quadrant of the East Zone and is significantly lower in the West Zone. Because of the lower elevation in the West Zone, the water table is significantly higher, and, as a result, one of the reservoir sites contains a set of sump pumps to continuously pump down the water table. The reservoir is located below grade at a sports park, with a tennis court on top of the reservoir. The sump pumps operate an average of 16–20 hours each day. In the event of the sump pumps being completely unavailable, the water table would lift the below-grade reservoir out of the ground.

There are three sump pumps configured in a Lead-Lag1-Lag2 configuration controlled by a Milltronics HydroRanger. The control is intentionally hardwired through the level controller and is completely independent of the city's PLC. In this way, in the event of a PLC failure, SCADA failure, or cyberattack on the control system, the level controller will continue to operate the sump pumps as a standalone system. The city's PLC only receives feedback from the sump pump system, consisting of failure of the level controller and failure of each of the sump pumps.

The reservoir site currently is not equipped with an emergency generator. The availability of electrical power is critical to keep the sump pumps running. To ensure the availability of the sump pumps, the city has installed a quick-connect electrical hookup, which is connected to an automatic transfer switch (ATS). The city has a service contract with a vendor to obtain a generator on a trailer with minimal downtime to the sump pumps.

Should the sump pumps be unavailable for longer than an hour or two, it would be a very bad day for the city. The city has taken measures to ensure this doesn't happen, namely,

- control the sump pumps through hardwired control with no provisions for control through the PLC; and
- ensure technology, capabilities, and service level agreements are in place to procure emergency power in a timely manner.

CASE STUDY #8: MAINTAINING LEVELS OF SERVICE FOR SEWER LIFT STATIONS

The city's sanitation department consists of three lift stations and approximately 10 SmartCovers to monitor sewage flow. Sewage flows into each lift station through an influent main, where it fills a wet well. Once the wet well reaches its high-level set point, the lead pump turns on. The pump continues to run until the level in the wet well reaches the stop-level set point. If the lead pump is running and the level continues to rise, the lag pump is called to run. Sewage is pumped out of the lift station through a forced main to Orange County Sanitation District for wastewater treatment.

Because there is no valving to isolate the lift stations and there are no provisions to reroute influent flow elsewhere, the pumps at each lift station are critical. Recognizing the criticality of each pump, the city has hardwired the control of both pumps at each lift station to a set of redundant Siemens HydroRanger 200 level controllers.

Unlike the reservoir case study previously discussed, the pumps in this case study are hardwired to the PLC for level control in addition to being hardwired to the level controllers in the field. Under normal operations, the PLC controls the pumps. However, in the event of a PLC failure, SCADA failure, or cyberattack on the control system, a selector switch on the control panel will assign control of the pumps to the level controller. When operating in this mode, the level controller will operate the pumps independently of the PLC.

Another critical concern is the availability of emergency power. The city has installed generators and ATS at each of the lift stations. Operations staff inspect the generators and ATS at each site daily, ensuring that there is sufficient fuel and that there are no issues or alarms with the emergency power equipment. Operations staff also regularly conduct exercises to confirm that the generators and ATS are indeed functional so that when they are called on during an emergency, standby power will be available instantaneously.

Similar to the sump pumps in the previous case study, the lift stations in this case study are all critical pieces of equipment. Should the pumps be unavailable for more than an hour, it would be a very bad day for the city's sanitation department. The city has taken similar measures to ensure this doesn't happen, namely,

- having hardwired control of the pumps through redundant level controllers should there be a failure of the PLC, SCADA system, or a cyberattack on the process control system;
- ensuring emergency power is installed and functional; and
- conducting regular tests to confirm that failover to emergency power happens instantaneously to minimize disruption to operations.

CASE STUDY #9: DETAILED CCE ASSESSMENT

The case study provided in Appendix A provides a detailed discussion of a consequence-driven, cyber-informed engineering (CCE)/critical function assurance (CFA) assessment of a water utility. In the fictional case study, the CCE/CFA assessment team completed a full assessment of a water system, working through all four phases.

It is likely that the assessment included here is representative of numerous water and wastewater utilities in the United States based on the application of technology, operational assumptions, procurement of outside support for operations technology (OT) implementation and maintenance, and engineering of the infrastructure.

For those utilities that are interested in a detailed CCE/CFA assessment, this can provide a road map. It can also be combined with other case study resources, such as those found in the *Countering Cyber Sabotage: Introducing Consequence-Driven, Cyber-Informed Engineering (CCE)* book (Bochman and Freeman 2021) and on the Idaho National Laboratory website.

REFERENCES

Bochman, A.A. and S. Freeman. 2021. *Countering Cyber Sabotage: Introducing Consequence-Driven, Cyber-Informed Engineering (CCE)*. Boca Raton, Fla.: CRC Press.

Wright, V. L., J. P. Meng, R. S. Anderson, et al. 2023. *Cyber-Informed Engineering Implementation Guide*. INL/RPT-23-74072. Idaho Falls, Idaho: Idaho National Laboratory. https://www.osti.gov/biblio/1995796 (accessed April 15, 2025.

Chapter 5

CIE Case Studies

CASE STUDY #10: REMOTE US WATER UTILITY AND SCADA IN THE CLOUD

A remote US water utility was struggling to maintain consistent operation of a critical pump station that feeds customers up a ridgeline in one of its regions (see Figure 5-1). The control system at the pump station was well past its useful life, and operators were required to address alarm conditions in person every day. Because of the traffic in this state, this required approximately ½ a full-time equivalent (FTE) to address alarm conditions at just this one pump station.

The remote US Water Utility decided to implement a supervisory control and data acquisition (SCADA) in the Cloud solution at the pump station to allow for both monitoring and control. Recognizing that this could free up ½ an FTE worth of people-time, an advanced water sector engineering firm supporting the utility decided the Cloud solution could be valuable. However, the project team understood the risks posed to the utility by potentially allowing for control through the Cloud.

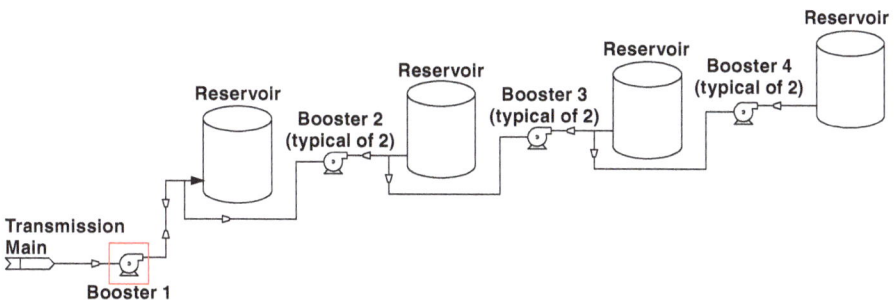

The pump station is responsible for drawing water from a trunk main and pumping it up to a series of pump stations and reservoirs along a ridge of residential customers.

Figure 5-1   The critical pump station is shown in the red box

As a condition of implementing SCADA in the Cloud, the western US engineering firm conducted a consequence-driven, cyber-informed engineering (CCE) assessment to complete a cyber-informed engineering (CIE) review of the pump station. This was done to ensure a resilient implementation and limit the potential consequences of a malicious adversary taking control of the pump station. At the time, the CCE methodology was well-defined, while CIE was not. The CIE Implementation Guide was still several years away from being published. In addition, the engineering firm's staff were well trained in applying CCE at that time.

The firm conducted the CIE review. It started with the definition of the worst reasonable consequences.

Proposed Facility Control System Changes

Current operations do not allow visibility through SCADA for operations on the ridge, and staff must respond in person to any operational disruption. As an interim solution to improve the visibility and control of a pump at the facility, a SCADA in the Cloud solution was proposed. The Cloud solution allows for remote monitoring and control through the Cloud-computing platform (CCP) of a multinational technology company. Based on the utility's experience with the specifically designed interim solution, known as "Emergency SCADA," for monitoring the facility, operations personnel had requested control capabilities in addition to monitoring. The capabilities would be applied to one of the two pumps at the pump station. Based on the Cloud SCADA service model and the utility's experience to date, the interim solution is expected to provide reliable monitoring and control capabilities. However, implementing a Cloud solution carries substantial cyber risk with it. The following is a summary of a CIE/CCE assessment of the cyber risk and recommended mitigations/protections that the utility should implement before connecting to the Cloud SCADA platform.

CIE/CCE Approach

The project team chose to follow the four-phase CCE process shown in Figure 5-2.

One of the concerns heard from an organization that sets out to complete a CIE/CCE assessment is that it sounds like a resource-intensive effort. The project team was careful to constrain this assessment in a manner that allowed for a quick delivery and concise analysis and set of conclusions. This assessment had a narrow scope of pump station operations up the ridge. Relative to the utility's full-scale operations, this is a very small portion of their system; however, this may inform ongoing upgrades in other portions of their's infrastructure.

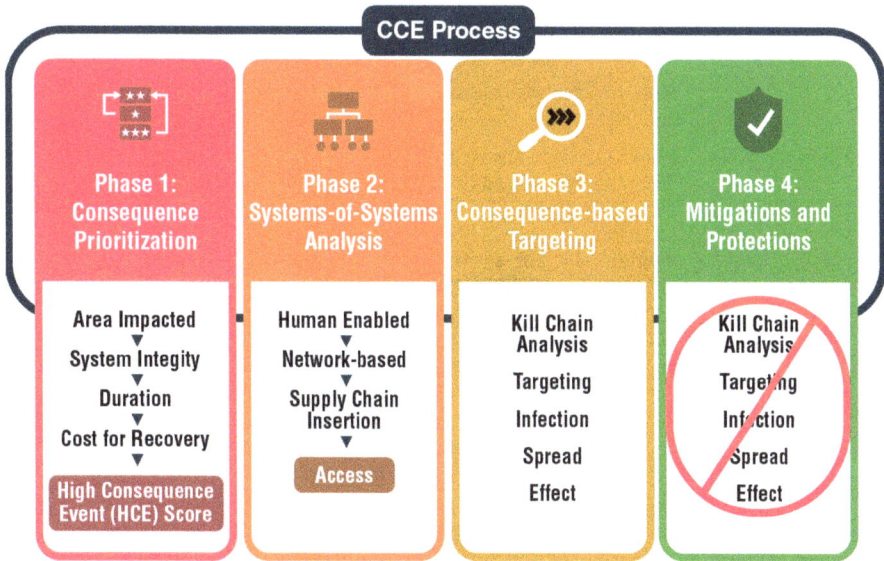

Source: Used with permission from BEA/INL

Figure 5-2   CCE methodology

Assessment Assumptions

The scope of the Cloud SCADA technology deployment included both monitoring and control. The deployment included a single field unit at a single station that is started/stopped remotely through the Cloud SCADA platform.

CCE Phase 1: Consequence Prioritization

The first phase of CCE includes a detailed assessment of negative events that could affect the pump station's critical functions. Specific to this assessment, the critical function is the safe delivery of water up the ridge for customers to use. The following scenarios were identified as those with the highest potential consequences.

Phase 1 assumptions.
- The adversary has logical and physical access.
- The adversary is knowledgeable of critical equipment, processes, and how to affect the system.
- The adversary is well resourced and capable.

To begin the analysis, the project team determined the objective, scope, and boundary conditions (Table 5-1).

Based on the objective, scope, and boundary conditions developed, several cyber events were identified and evaluated as summarized in Table 5-2.

Table 5-1   The objective, scope, and boundary conditions for the CIE/CCE review

Objective	Disrupt drinking water service on the ridge for more than 24 h.
Scope	Meet the objective by sabotaging the booster pumps controlled by the Cloud-based system.
Boundary Conditions	Disrupt drinking water service on the ridge by sabotaging the booster pumps controlled by the Cloud-based system, causing an outage for more than 24 h.

Table 5-2   Ranking of potential cyber-events

Cyber Event	Cyber-Event Description	Severity Ranking (1 = Most Severe)
Overfill reservoir #2	The adversary gains access to the pump station controls through the Cloud platform. Malicious modifications cause the pump station pumps to overfill the next reservoir up the ridge from the Cloud implementation. Display on the Cloud platform appears normal.	3
Start/stop pump multiple times to cause physical damage	The adversary gains access to the pump station controls through the Cloud platform. Malicious modifications cause the pumps to start/stop rapidly, damaging the pump and causing service disruptions on the ridge. Display on the Cloud platform appears normal.	1
Drain reservoir #1	The adversary gains access to the pump station controls through the Cloud platform. Malicious modifications cause the pumps at the station with the Cloud implementation to drain the reservoir, causing service disruptions on the ridge. Display on the Cloud platform appears normal.	2

Based on the evaluation completed and supply chain concerns during the time of the assessment (mid-COVID-19 pandemic), a scenario resulting in damage to the pump was the greatest concern.

CCE Phase 2: System-of-Systems Analysis

Based on the training offered by Idaho National Laboratory (INL) (https://inl.gov/national-security/cce/), phase 2 can represent a large portion of any CCE assessment. This is the portion of the project in which all of the various systems are defined, and the project team builds an understanding of the current cybersecurity protections and resilience capabilities in the system.

In addition, every organization has numerous enabling functions that support the delivery of the critical function. For this assessment, the critical function, as noted previously, is the safe delivery of water up the ridge for customers to use.

A general MindMap graphic showing common critical functions and enabling functions for a water utility is presented in Figure 5-3.

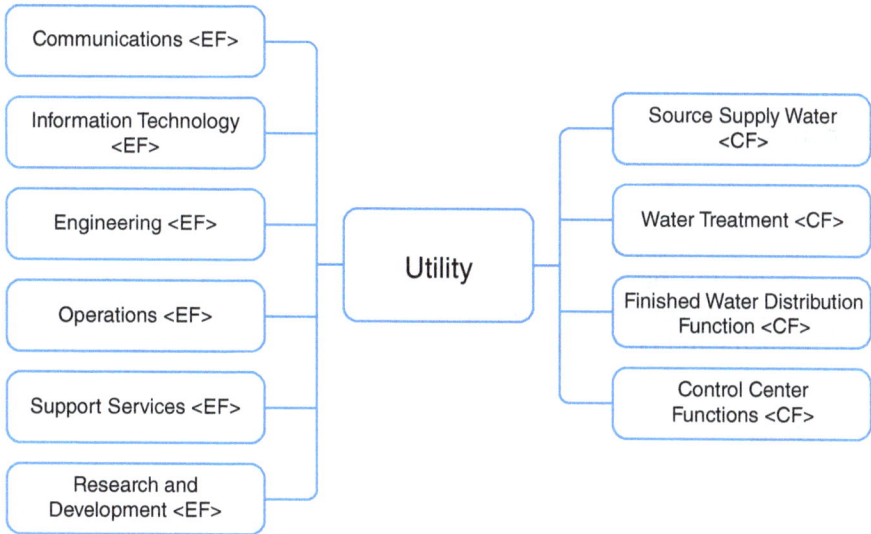

Figure 5-3   A general MindMap graphic showing common critical functions and enabling functions for a water utility

Enabling functions for this assessment included
- Utility staff;
- Cloud platform staff;
- the Cloud platform and CCP; and
- the cellular network.

Because the Cloud platform and CCP are largely proprietary, the project team didn't have access to detailed design and operation documentation on these. Although the CCP was not engaged as part of this assessment, the Cloud platform vendor was happy to support the assessment. Based on the vendor's expertise, the project team engaged them to determine how an attack might occur.

CCE Phase 3: Consequence-Based Targeting

In consequence-based targeting, the project team develops a targeting (e.g., attack) scenario by which the highest-ranked cyber event could be made to occur by an adversary. Although this can take the form of many drawings and details, the project team felt that a high-level summary was more appropriate. The adversary pathway was broadly defined as:

Public Internet → CCP → Cloud Platform → Pump Station Physical Assets

Given the proprietary nature of the CCP and the Cloud platform, the Cloud platform vendor provided insights into how an insider at the vendor would carry out an attack. The project team asked the vendor to take this perspective because it leveraged the "insider" understanding of the platform. Numerous excellent cybersecurity controls were implemented; however, a targeting scenario was developed.

CCE Phase 4: Mitigations and Protections

During phase 4 of the project, the team determined which mitigations and protections would be implemented to ensure the resilience of the system.

It is always best to prioritize protections because they take the attack scenario off the table; however, these may not always be possible. So, mitigations are always considered as well. Mitigations make the attack more difficult and potentially reduce the value of the attack scenario for the attacker.

Protections

The following protections were recommended to limit the risk of a cyber incident:
- Physical protections
  - Timer relay—This prevents someone from turning the pump on and off, potentially damaging the pump and requiring replacement.
- There are two pumps onsite, which only connect one of the pumps to the Cloud SCADA field unit. A corresponding mitigation is to establish a contingency plan to switch pump operations from the Cloud-connected pump to the non-Cloud-connected pump (business-as-usual operations scenario) as needed.

Mitigations

The following mitigations were identified:
- Implement and maintain all of the cyber-hygiene controls currently in place by the Cloud vendor.
- Implement access control and a user hierarchy.
- Limit the number of users with access and control capabilities.
- Implement logging to records and alarm if unauthorized access is attempted or occurs.
- Anonymize screens to the extent possible—this was a recommendation from utility staff. Because they had only one facility on the Cloud platform, it was simple to obfuscate the pump station name to increase the difficulty of an attack.

Conclusion

This case study is one of the clearest available examples of how CIE/CCE may be used to quickly and decisively improve the implementation of technology. In addition, it clearly demonstrates that CIE/CCE can have the desired effect on the project team and the resilience of the utility.

CASE STUDY #11: CITY IN NORTHERN CALIFORNIA– CIE DESIGN GUIDELINES

Following on from the A Day Without SCADA exercise in a Northern Californian city, leadership recognized that they needed to embed some of these principles within their design guidelines so all engineering projects for the city's water and wastewater systems included CIE principles. To do this, they drafted a design guideline preamble. The draft version is as follows:

To identify engineering and staff capability improvements, the city's Department of Utilities conducted a tabletop exercise (TTX) in October 2020 simulating a loss of automation. During the TTX, numerous existing positive engineering practices and improvements to those practices were identified. In response to this TTX and the changing cyber-risk landscape, the city has adopted a revised strategy for the design of water treatment and pumping facilities. This new strategy is consistent with the principles and practices of INL's CIE. At a high level, this revised strategy requires design engineers to:

- accommodate operations in the absence of automation (meaning unreliable or unavailable human–machine interface, programmable logic circuits (PLCs), historian, or communications) for operation and monitoring of pumping and treatment; and
- include protective measures to prevent damage from manipulation of the control system by a malicious cyberattack.

Therefore, engineered systems must

- operate at the required service levels in the absence of automation;
- be designed to minimize the number of staff required to operate a process, facility, or asset in manual operations;
- be designed to minimize the complexity of manual operations;
- include analog instrumentation at critical system monitoring points;
- include contingency planning for
  - ○ safe transition from automated to manual operation,
  - ○ safe operation in the absence of automation, and
  - ○ safe transition from manual operation to automated operation; and
- include protective measures to prevent damage to physical assets from manipulation of the control system during a malicious cyberattack (e.g., backspin timer relays).

The city will conduct CIE-based design reviews as appropriate during the design process to ensure these objectives will be achieved.

CASE STUDY #12: SCMWD'S JOURNEY TO CIE

In October 2024, the staff of a Southern California municipal water district ("SCMWD") participated in a CIE workshop to build awareness amongst engineering and technology staff. Staff are always overbooked, and spending several hours in a non-project-related workshop can be a tall ask. Because of this, staff were questioning the value of spending the time in this workshop compared with their other day-to-day responsibilities. One staff member, a senior electrical engineer, was particularly skeptical of the workshop. However, during the workshop, he had an epiphany.

The senior engineer had been concerned about the issues that CIE has been aiming to resolve for many years as he observed the increasing digitization and connectivity within the utility's water and wastewater systems. He just didn't have a conceptual framework to effectively communicate with his leadership on the topic. This workshop gave him the concepts and associated terminology to communicate with both leadership and his fellow engineering staff.

In the wake of this workshop, the staff decided to directly apply CIE within an ongoing water treatment facility design process. The CIE team conducted reviews at both the 60% and 90% design milestones. The high-level findings are summarized as follows:

- 60% design milestone
  - Milestone background: At the 60% design milestone, only high-level process-mechanical, electrical, and instrumentation and control (I&C) design elements have been developed.
  - The review team made approximately 10 CIE-related comments.
- 90% design milestone
  - Milestone background: At the 90% design milestone, most of the design is drafted, including process-mechanical, electrical, and I&C.
  - The review team made approximately 100 CIE-related comments. The number of comments was directly correlated with the increased definition of the design.

Figure 5-4 shows the describes the maturity of design elements by milestone.

Although some of the comments were more important than others, the utility and design engineer found that the comments were impactful and resulted in potential cost changes to both the design and construction of the water plant.

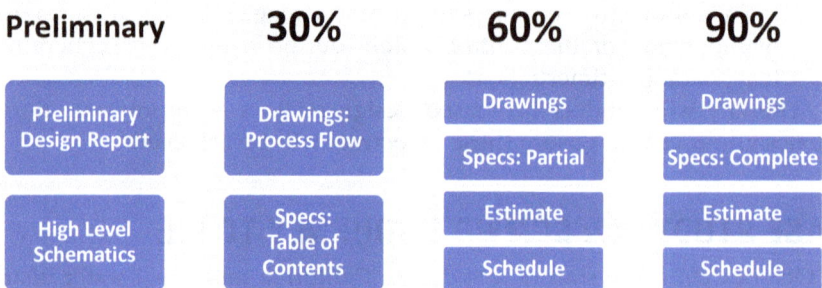

Figure 5-4   The maturity of design elements by milestone

Review workshops and meetings were held throughout the project. One unexpected midproject outcome was the inclusion of other utility staff in the meetings. They were invited so that they could observe CIE in action and bring the practices to their projects.

The CIE review team consisted of engineers and operations technology (OT) specialists, including a white-hat hacker. The CIE review team learned several important lessons throughout this project as well. These can be summarized in two terms that are helpful for project team members to

The first is *cyber-enabled failure mode*—a failure mode intentionally caused in a system via cyber means. Engineers are familiar with the concept of failure modes. This familiarity is built on centuries of experience building and maintaining infrastructure systems. The difference for CIE is that an attacker could be intentionally trying to cause a failure mode. Many engineers do not have the defender's mindset. This concept helps them form that up and reduce the potential for a failure of imagination that all engineering teams could be subject to.

The second is *commander's intent*—for those who served in the military, this may be a familiar term. For civilian engineers conducting the CIE review, this became an important novel organizational principle. Early in the CIE review project, the engineers did what they did best. They provided engineering comments based on the design. What was lacking was a CIE-specific organization to the comments. Commander's intent provided the organizational scheme that supported with evaluation and communication of cyber-enabled failure modes.

Implementing this concept was best done through targeting meetings. One of the team members established himself as the mission commander and identified pumping and process equipment that, if manipulated, could provide the greatest effect on the utility's critical functions. He then instructed the rest of the CIE review team to confirm that those pumping and process areas could be targeted and would have the desired effects. In several cases, the team determined that some of the commander's targets and desired cyber-enabled failure modes were valid, while others were not. With the implementation of this organizational structure, it provided an easy way to communicate within the utility and with the design engineer.

Figure 5-5 shows the timeline for this utility to complete the CIE review. It was fully integrated into the standard design process.

The CIE reviews were not scoped before the 60% project. Therefore, the preliminary design report (PDR) and 30% design review are not included in Figure 5-5. CIE-related comments at these stages are limited. However, these can be approached to set up the design team for success in the later stages.

SCMWD has chosen to develop CIE design guidelines to provide guidance to all future infrastructure projects at the utility. To codify CIE within the engineering culture, guidelines were created to direct SCMWD engineering teams on how to establish a CIE team for their project and how to go through the review process while addressing the greatest risk items and capturing lessons learned.

Figure 5-5 Timeline for the utility to complete the CIE review

CASE STUDY #13: NORTHWESTERN WASHINGTON WATER AND WASTEWATER DISTRICT—CIE SERVICES

A water and wastewater district in Northwestern Washington contracted with the western US engineering firm to develop a SCADA master plan (SMP). Although the traditional SMP is focused on what technology the utility will implement, this SMP includes extensive discussions of resilience. This included a consequence-driven, cyber-informed engineering (CCE)/critical function assurance (CFA) training; A Day Without SCADA exercise; and a CIE design review of a critical pump station.

The district was already identifying and implementing CIE-related protections within many of its engineering projects; however, there was a great deal of concern about retaining these practices as historical best practices to address current and future cyber risks. Several engineers who were long-time district employees recognized this need and championed this project.

Training and A Day Without SCADA Exercise

To start the project and build awareness of the technical concepts, the firm led CIE/CCE/CFA training. This was done to help the district understand that the primary objective of the series of training was not about cybersecurity. It was about ensuring that the infrastructure was designed in a manner that allowed the district to achieve its critical functions during a cyber incident.

During the training and exercise, it was noted that the district had intuitively retained many historical engineering best practices that would prevent damage from a cyberattack. For example, physical electromechanical protective relays were still required by the senior engineering staff in all designs.

In addition, it was noted during the workshops that in the engineering design process, although inclusive of staff from numerous disciplines across the organization (e.g., operations), many of the important feedback loops were not formalized.

One immediate challenge that the district faced was the turnover of the senior engineering staff. This resulted in a concern that these best practices would be lost and result in long-term consequences for the utility.

The CIE/CCE/CFA training was key to helping the district document the engineering process and enshrine many of the engineering design best practices. In addition, it set the stage for the CIE review of a large pump station that was currently in the design process.

CIE Review

The western US engineering firm conducted CIE reviews of the pump station design, delivered via design-bid-build project delivery, at the following design milestones:
- Preliminary design report/30% design
- 60% design
- 90% design (pending as of the writing of this book)

Preliminary Design Review/30% Design

At this stage, there is limited opportunity to address cybersecurity or CIE in a detailed manner; however, during this review, it was noted that the design did not include a simple mention of designing to a specific cybersecurity standard either new or already adopted as the district's standard.

60% Design

At the 60% design review milestone, the project team completed a CIE review. The comments continued to be relatively high level given the nature of the design deliverables provided at this milestone with limited electrical and control system content. However, several comments were provided. These focused on the following:
- Ensuring the potential for manual operations.
- Improving the usage of more sophisticated network traffic monitoring solutions. This includes training staff on how to leverage the capabilities for day-to-day operations and incident response.
- The design incorporated a vendor-provided solution. Based on the information provided by the vendor, there were numerous questions on the cybersecurity controls and operations capabilities to operate in the absence of automation.

90% Design

During this review, it was discovered that there was very little information relative to the communications and control systems available for the CIE review. The design engineer (via the contract specifications) had delegated design of the key control system components to a "systems integrator" who was identified as a second-tier subcontractor to the construction contractor.

The systems integrator was going to be selected by the construction contractor at a later date and likely to be the low-bidder for that portion of the project. Unfortunately, the delegation of design to systems integrators is a common practice among engineers who serve the water sector.

The contract documents did not include any requirements for the systems integrator to have any knowledge on CIE, cybersecurity, or networking. Their focus was to be on PLC programming and SCADA software configuration.

This CIE review resulted in the District making a significant change in the direction they provide to their design engineers relative to the delectation of design elements to the contractor. This includes such cybersecurity practices as secure networking and firewalls.

The district is moving forward with making significant changes to how they procure services and implement engineering projects in the future. These changes include updating guidance for consulting engineers and the selection process for engineering services.

CASE STUDY #14: SCMWD PROCUREMENT

In a first of its kind, SCMWD provided procurement language within a Request for Proposal (RFP). This RFP was circulated to preferred consultants. It is for the final design of a large pumping facility within the district's system.

The RFP language developed is quoted as follows:

> *"Cybersecurity shall be considered foundational to the design. Consultant shall utilize cyber-informed engineering (CIE) and consequence-driven cyber-informed engineering (CCE) to identify areas that may be at risk to cyberattack and mitigate risk through purposeful design to reduce or remove risk by enhancing resilience. Consultant shall conduct separate focused cybersecurity workshops prior to the 60% and 90% design submittals. Workshops shall discuss risk identification, consequence, prioritization, interdependencies, controls, design simplification, and other mitigation measures. Mitigation measures shall be incorporated into the design. Workshops may be combined with other meetings where appropriate, subject to District approval."*

This work has not yet been completed. SCMWD has embraced CIE perhaps more than any other asset owner in the United States. When procured and implemented from the beginning of a project like this, it ensures that CIE will be included throughout the engineering life cycle.

CASE STUDY #15: MAINTAINING THE DEFENDER'S ADVANTAGE

One of the tenets of CIE/CCE is that the adversary will always know at least as much about the individual technologies deployed in an OT environment, maybe even more than the owner and the manufacturer. But they won't always know what technology is deployed or how the technology is deployed and configured to deliver a critical function. Protecting a utility's information should be considered paramount. Successfully protecting this information results in the defender's advantage. This is maximized when the utility prevents any information from entering the public domain. While this might not be practical, it provides the appropriate starting point for any organization. The analogy used in INL's CCE training is to imagine your adversary is in a dark room (i.e., your OT environment), and as you provide them with more information (e.g., network architecture, hardware deployed, software deployed), it is like turning the light on so they can navigate any obstacles you may have set up.

Because most utilities are public agencies, public procurement requirements apply. Generally, the intent of these is to drive fair procurement practices. However, they also can result in publicizing the specific make and model of hardware and software that may be in use at the utility, providing an adversary with an excellent starting point.

Other utilities, like one of our best clients, develop general engineering standards and post them online. This allows for the standardization of engineering practices for many projects and saves staff time with coordination. Unfortunately, as of this writing, this includes the make and model of PLCs and networking devices.

We have observed our clients successfully work with their procurement colleagues to constrain the amount of information provided to prospective contractors. There are various ways to do this. Our preferred method is to establish rigorous information protection and transfer methods with a limited number of trusted contractors, to the extent outside support is needed. The Department of Defense has the Cybersecurity Maturity Model Certification Program designed to protect unclassified information. Perhaps this is something that the water sector should work toward to help ensure the protection of utilities' information.

Chapter 6

Bringing It All Together: How Do We Apply CIE to Our Water/Wastewater Systems?

So, what are some best practices for applying cyber-informed engineering (CIE) to our water systems? As consulting engineers, our only clear answer is—"it depends." Our water and wastewater systems are in a process of continual renewal. The early 2020s have seen an increased rate of investments in water infrastructure. As demonstrated by the aforementioned case studies, CIE may be applied at any phase of the asset life cycle. We have found that, invariably, utilities are already doing something right when it comes to CIE, whether intentionally or not. Regardless, this "something right" provides the core to build additional best practices. The following sections provide additional context on applying CIE for infrastructure sign projects and existing infrastructure systems, building on the case studies presented previously.

FOR INFRASTRUCTURE IN DESIGN

Conducting CIE reviews is indispensable for any project in design. Our experience is that bringing awareness of CIE to the design team often results in both subtle and nonsubtle shifts in the final design of the infrastructure but also in how future projects are executed.

Some of the subtle shifts we have observed include:

- Engage cybersecurity staff earlier in the design. Of course, the staff are often from operations technology (OT)/information technology (IT) groups. They will struggle to maximize the application of CIE because their purview often extends only to the router. They will likely not support the evaluation consequences and the identification of engineering controls. So, connecting the OT/IT staff to the engineers in a discussion allows for a greater understanding of risk than otherwise.

55

- Another observation is that discussions of CIE help the IT/OT staff relax. We think that this is because an assessment that assumes the adversary is inside the system and "what do we do now?" helps everyone understand the organizational capabilities and limitations. This also helps the IT/OT staff express how their greatest concerns (i.e., penetration through the defenses into the control system) are being addressed in a structured manner.
- Considering CIE throughout the design educates staff and contractors. They will take these lessons learned to improve the cyber resilience of future projects.

The best practice for projects in design is to complete CIE reviews at each of the design milestones (preliminary design, 30%, 60%, and 90%) while changes can still be made to the design. Building security and resilience into infrastructure in the design phase is generally considered to be less expensive than "bolting it on" later in the life cycle.

CIE CHECKLISTS

Many people around the sector have requested CIE checklists. This is a bit of an easy button, but also, engineers are known for their desire for completeness and rigor. It is a challenge to measure those two things without something to measure against. Checklists offer that in a familiar format commonly used across the sector.

A word of caution—checklists are not the intent of CIE. This is one of the reasons why the CIE Implementation Guide starts with 1,152 questions. The intent of CIE is to change how we engineer and think about the infrastructure systems that we design and rely on every day. There is ambiguity in that intent that we will have to sort out through the application of CIE. We fully expect that there will be CIE checklists someday relatively soon. These will best cover simpler facilities like small booster pump stations or lift stations.

In the meantime, while checklists are developed, how we have conveyed the benefits of CIE to our clients is best captured in three key questions:

- How are your systems engineered to be operated without automation? When the systems were designed and constructed, were manual operations capabilities built in? If they were not practical (e.g., for advanced forms of treatment) what contingencies were designed and implemented?
- How are your staff trained to operate your systems without automation?
- How are your systems engineered with cyber-physical protections?

Regardless of the approach taken, it is critical to prevent a "failure of imagination" (Wikipedia Contributors 2025). This term has been used in the wake of various national tragedies or emergencies such as 9/11. An engineering team that succumbs to a failure of imagination would fail to identify cyberattack scenarios that could be prevented through improved engineering and operations practices.

Certainly, for larger and/or highly complex infrastructure projects like new water treatment plants, we propose that a formal CIE team be formed both to achieve rigor and to address the complexity inherent in modern-day critical infrastructure.

For those who thrive on robust and lengthy checklists, there is hope. INL is currently leading the development of a CIE maturity model. Inherent to this is the development of behaviors and characteristics that an organization can evaluate itself against. In time, any utility will be able to determine where it sit on the maturity model. Depending on the outcomes, the utility will be able to establish a clear path to achieve the next level of CIE maturity.

FOR EXISTING INFRASTRUCTURE SYSTEMS

It would be magnificent if every utility could set aside the people, time, and budget to conduct a broad CIE/consequence-driven, cyber-informed engineering (CCE) assessment of their organization today; however, that is unlikely for the vast majority of entities. That's OK. There are many starting points articulated in the previous section.

What we have found to be a logical process for a utility to go through to begin adopting CIE within their engineering practice is shown in Figure 6-1. Although the figure is too high level to show it, there will inevitably be feedback loops throughout the process. For example, one of the approaches we advocate for is to have the equivalent of health and safety tailgate discussions to discuss manual operations. This allows for quick process or pumping station–specific discussions that operations staff can complete without a more time-intensive endeavor like a tabletop exercise (TTX).

A Day Without SCADA → CCE/CFA → CIE

Another best practice that has emerged is that to get to CIE, a utility should go through an A Day Without Supervisory Control and Data Acquisition (SCADA) exercise and CCE/critical function assurance (CFA) assessment first and in that order. Utilities, for good reason, put a great deal of trust in their operations staff. So, when we have presented CIE to utilities, we often hear that "the operations team will handle it." Although this may be true, it is a broad assumption. Many organizations haven't had to deal with even a SCADA outage that provides the operations team with an opportunity to explore operating the system in the absence of automation.

Conducting an A Day Without SCADA exercise first allows for an evaluation of the assumption that the operations team can handle it. Getting that question out of the way allows the team to move on to a CCE/CFA assessment. These can be big or small in scale. Going through the full four phases of a CCE/CFA assessment can be daunting to a resource-strapped organization. We have found that even having a workshop on the topic can help answer some of the questions regarding cyber-physical controls in place, response and recovery capabilities, and assumptions design engineers have already made about operating through a sophisticated cyberattack.

Develop Awareness

It is likely that the utility is just learning about CIE. Building awareness is the first step. This can be done though a meeting or a workshop. Generally, an hour isn't enough because the topic is compelling to most utilities.

Conduct an A Day Without SCADA® Exercise

We often hear in the awareness-building discussion that the utility is not worried about it because their operators "have it covered." The operations staff are talented and knowledgeable in their field of practice. However, this assumption must be tested and the best way to do it is through a tabletop exercise that creates a positive learning environment for the whole utility team. This allows the team to ask and answer two questions. First, are the systems engineered in a manner so that they can be operated without automation? Second, are operations staff appropriately trained to operate the systems in the absence of automation?

Conduct a CCE/CFA Assessment

The next step is to conduct a CCE assessment. As noted in the case studies in this book, this doesn't have to be a large project. It can address an existing concern within the existing system or review a system in design. This can be successfully done in hours to days and be impactful to the staff involved.

This answers a third question: What cyber-physical protections and mitigations does the utility have in place to ensure that, if a cyber-attacker is in the network with control capabilities, they cannot impact the critical functions of the organization. This explores what protections a utility has in place to ensure the worst cyber-days don't occur.

Implement CIE

As a next step, it is important to capture the lessons learned within CIE guidelines or similar design documentation to ensure that the organization creates and retains best practices.

Figure 6-1   The most effective pathway for the adoption of CIE at water and wastewater utilities

Once the utility has worked through a CCE/CFA assessment, the organization is most ready to move into CIE. To a large extent, staff are comfortable with the common assumptions between the approaches, are familiar with the terminology, and understand the benefits of CIE to their organization.

As of the writing of this text, the structure of CIE assessments is relatively nascent. We look forward to more practitioners providing their perspectives to the conversation. We have our experience, but that is limited to an engineer's perspective. Generally, this means reviewing the engineering guidelines, specifications, and standards. Then, we identify potential improvements and/or develop the first iteration of a CIE-informed version of these documents. However, CIE assessments can be much broader given the breadth of the principles of CIE. In time, we expect other portions of critical infrastructure organizations to engage heavily with CIE. For example, procurement staff must be brought on board to ensure *cyber-secure supply chain controls* (principle 9) are created and maintained, or executive leadership will need to be brought in to drive *organizational culture* (principle 12) change.

In late 2024, the authors began working with clients to enshrine CIE within their engineering practice. The following figure summarizes the pathway that we have found to be most effective in the adoption of CIE at water and wastewater utilities.

Of course, there will be other great ways to achieve sustainable implementations of CIE within organizations and we look forward to hearing those as utilities take on the challenge.

Risk and Resilience Assessment/Emergency Response Planning

The America's Water Infrastructure Act of 2018 (AWIA) Section 2013 created a requirement that water utilities serving more than 3,300 people shall
- conduct an assessment of the risks to, and resilience of, its system, and
- prepare or revise an emergency response plan (USEPA 2025).

AWWA Standard J100, Risk and Resilience Management of Water and Wastewater Systems, is the best practice for conducting a risk and resilience assessment (RRA) (AWWA 2021). AWWA J100 is consistent with international enterprise risk management (ERM) best practices such as ISO 31000. The J100 seven-step process is shown in Figure 6-2.

One approach to integrating CIE with the J100 process can be adapted from the CIE and ERM guidance document published by Idaho National Laboratory (INL)/Department of Energy (DOE). To support cross-sector critical infrastructure entities with the adoption of CIE, INL developed the document CIE and ERM guidance. This document presents an approach that a J100 or ERM practitioner can integrate CIE throughout an assessment project. The INL/DOE document includes a flowchart by which a CIE-centric process may be used within an RRA/ERM project. This has been adapted to align with AWWA J100 Standard and is included in Appendix B. Before using this flowchart, the reader should read the background provided in the INL/DOE document.

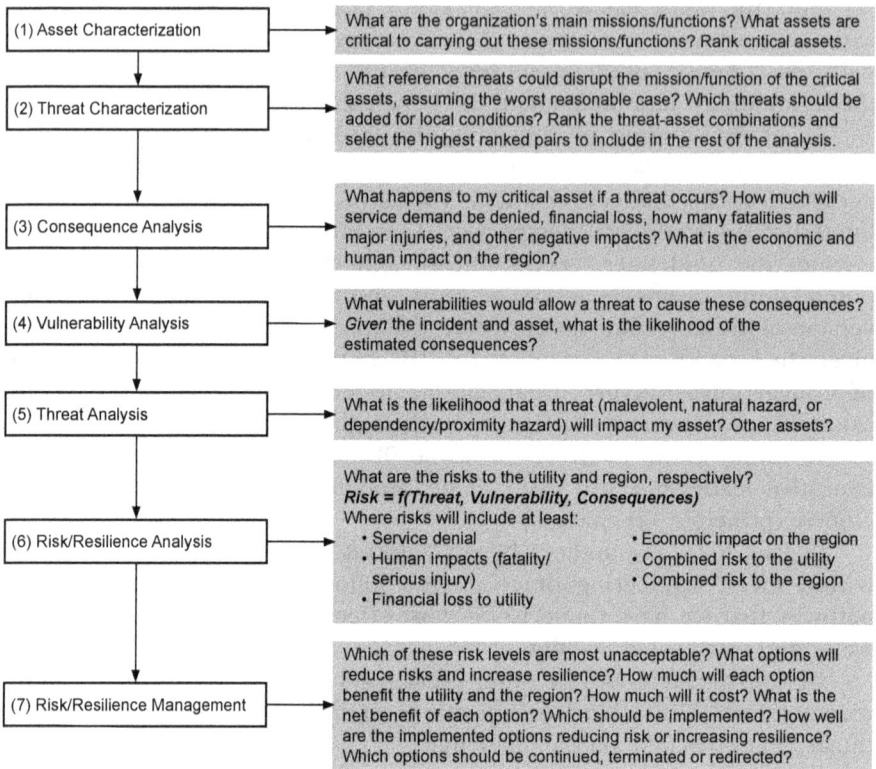

(1) Asset Characterization	What are the organization's main missions/functions? What assets are critical to carrying out these missions/functions? Rank critical assets.
(2) Threat Characterization	What reference threats could disrupt the mission/function of the critical assets, assuming the worst reasonable case? Which threats should be added for local conditions? Rank the threat-asset combinations and select the highest ranked pairs to include in the rest of the analysis.
(3) Consequence Analysis	What happens to my critical asset if a threat occurs? How much will service demand be denied, financial loss, how many fatalities and major injuries, and other negative impacts? What is the economic and human impact on the region?
(4) Vulnerability Analysis	What vulnerabilities would allow a threat to cause these consequences? *Given* the incident and asset, what is the likelihood of the estimated consequences?
(5) Threat Analysis	What is the likelihood that a threat (malevolent, natural hazard, or dependency/proximity hazard) will impact my asset? Other assets?
(6) Risk/Resilience Analysis	What are the risks to the utility and region, respectively? ***Risk = f(Threat, Vulnerability, Consequences)*** Where risks will include at least: • Service denial • Human impacts (fatality/serious injury) • Financial loss to utility • Economic impact on the region • Combined risk to the utility • Combined risk to the region
(7) Risk/Resilience Management	Which of these risk levels are most unacceptable? What options will reduce risks and increase resilience? How much will each option benefit the utility and the region? How much will it cost? What is the net benefit of each option? Which should be implemented? How well are the implemented options reducing risk or increasing resilience? Which options should be continued, terminated or redirected?

Source: AWWA 2021

Figure 6-2   The J100 seven-step process

A few notes on the adapted flowchart presented in Appendix B:
- The entire analysis is completed relative to the mission and critical functions of the organization. This is consistent between J100, CCE/CFA, and CIE. This is captured in step A.1.
- The analysis emphasizes characterizing the reliance on automation through numerous steps. This is captured in steps B.1, B.2, B.4, and C.1.
- To prevent the project team from moving on to later portions of the analysis without properly characterizing the reliance on automation, step D.1 provides an interim risk mitigation step to help staff improve their ability to operate the utility system in the absence of automation.
- A critical step of any CIE-related assessment is to go through the engineering controls "warehouse," as Ginger Wright from INL likes to say. This is shown in steps B.3 and C.3. It provides a critical step to understanding current protections before characterizing the potential consequences that an organization could realize during a cyberattack.
- Because both CCE/CFA and CIE roughly equate consequences with risk, step C.2 provides a step in which this characterization needs to

be complete by asking the question—are the consequences accept-able? All organizations carry some risk.
- The RRA team should pause at step D.2 and consider potential risk and resilience management strategies (i.e., improvement or risk mitigation projects) specific to mitigating types of consequences. Example types of consequences include
  ○ health and safety effects;
  ○ asset damage or loss;
  ○ financial effects;
  ○ environmental impacts;
  ○ economic effects;
  ○ public/customer confidence effects; and
  ○ loss of company confidential information.

In addition, Table 6-1 provides a mapping from the seven steps of J100 to A Day Without SCADA, CCE/CFA, and CIE. The notes provided should help a J100 practitioner integrate these different approaches and improve the cyber resilience of the utility under evaluation.

CIE is a natural addition to a utility's AWIA compliance efforts. Integrating CIE will help identify cybersecurity risk mitigations and protective elements (e.g., protective electromechanical relays) that are unlikely to emerge from a cyber-hygiene-centric cybersecurity assessment.

Table 6-1   AWWA J100, A Day Without SCADA, CCE/CFA, and CIE mapping

J100 Step	A Day Without SCADA	CCE/CFA	CIE
1—Asset characterization	Each of the three characterizes the importance of specific assets to the mission and critical functions of the organization.		
2—Threat characterization	The characteristics of the cyber threat are established as: • The adversary has logical and physical access. • The adversary is knowledgeable of critical equipment, processes, and how to affect the system. • The adversary is well resourced and capable.		
3—Consequence analysis	Each of the approaches helps identify the types and magnitude of consequences.		
4—Vulnerability analysis*	Based on the threat characterization, the vulnerability of the system to a cyber-attack is high based on cyber-hygiene only measures.		
5—Threat analysis	The annualized threat likelihood of one incident per year is established by guidance from federal agencies and real-world incidents.		
6—Risk and Resilience Analysis	Each of these approaches seeks to define the impacts/risks to the utility's mission and critical function. Each approach can inform the quantitative analysis.		
7—Risk and Resilience Management	Each of the approaches helps identify improvements (protections and mitigations).		

CCE—consequence-driven, cyber-informed engineering, CFA—critical function assurance, CIE—cyber-informed engineering, SCADA—supervisory control and data acquisition.

*Informing this J100 step is not a primary focus of these types of analyses.

CIE TEAM ROLES

One of the greatest lessons learned from our CIE projects is how to organize ourselves. As a team composed largely of engineers, we have all worked within the engineering design life cycle and associated milestones of

- preliminary design/schematic design;
- 30% preliminary design;
- 60% detailed design;
- 90% detailed design; and
- 100% bid-ready documents.

In addition, we have served as owner advisors and quality control reviewers on many projects. Because of this, our team naturally organized our CIE review team as a standard engineering project review.

In addition to the standard project team roles such as project manager, there are some CIE-specific roles that should articulated to deliver the best possible evaluation:

- *Team commander*—This person should set boundary conditions and expectations for the CIE evaluation. Depending on the nature of the system, this could include desired failure modes or downtimes for specific system components. For example, we have heard from multiple utilities that as their infrastructure ages, they become more concerned about a cyberattack resulting in potential overpressure conditions and damaging water mains. If a knowledgeable adversary is able to deduce the age of transmission mains through a document, such as a water master plan, those could be selected by the commander for targeting.
- *Subject matter experts*—This includes professionals from various engineering disciplines, control system specialists, operations specialists, and water quality specialists. This group provides perspectives on how the engineered system is protected from various failure modes and may be made to fail.
- *Targeter*—Simulating offensive cyber operations is a unique skill set all its own. This person has to be steeped in cyberattacker tactics, techniques, and procedures. One unique characteristic that this person should have is the inherent curiosity in how to "break things." In this case, that means circumventing cybersecurity and physical security controls and misusing an engineered system to achieve the attacking team's mission. Of course, this is a TTX only.

We learned the concept of a Tiger Team (Wikipedia Contributors 2025) from INL staff, and that emerged as an excellent reference model for how to organize a team. The genesis of this team formation is described previously in Case Study #13. This model also builds on the CCE Teams that INL articulates in ACCELERATE and Workforce Development training courses. More information on these courses may be found at https://inl.gov/national-security/cce/.

TECHNOLOGY TRENDS

Technology in a broad sense is changing so rapidly that no single organization or framework can address the ever-evolving cyber risks. Two technology trends in the water sector are of particular interest when discussing CIE: digital twins and artificial intelligence.

Digital Twins

Digital twins are a digital, dynamic system of real-world entities and their behaviors using models with static and dynamic data that enable insights and interactions to drive actionable and optimized outcomes (AWWA 2025).

A true digital twin has enormous potential to help water utilities improve system operation and maintenance. Of course, digital twins, like all things digital, are subject to being misused.

Imagine that your adversary has a real-world, near-real-time model of your infrastructure. How invaluable would that be? We hope for many readers, that makes you want to cancel the digital twin development project that you just kicked off. However, we hope you don't.

Like many things in life, and certainly in matters of security, there is a tradeoff between convenience and security. Build the digital twin and protect it accordingly, but assume that it will be comprised at some point. There are many great cybersecurity practices to help ensure that the model and associated data are held closely. It is a matter of building defense-in-depth around the model. There is amazing potential for this technology to improve operations, energy usage, and asset life for water and wastewater systems.

Finally, apply the principles of CIE to digital twin implementation, especially as these systems approach integration with the physical assets they mirror, including water distribution and wastewater collection systems.

Artificial Intelligence

Artificial intelligence (AI) is a generic term for "technology that enables computers and machines to simulate human learning, comprehension, problem solving, decision making, creativity and autonomy" (Stryker and Kavlakoglu 2024).

As a technology-focused society, we have been designing, using, and implementing AI since the 1950s (Figure 6-3). All of the automation systems that have been designed as part of our infrastructure are effectively AI because they have the ability to learn from historical data and can mimic certain human brain functions.

The latest iteration of AI, generative AI, has captured the attention of the world for both its potential benefits and dangers. Time will tell how this technology continues to take shape. We can expect that the technology will continue to develop and will likely start to have access to physical control of equipment. Once this transition happens, we will effectively have another potential "insider threat."

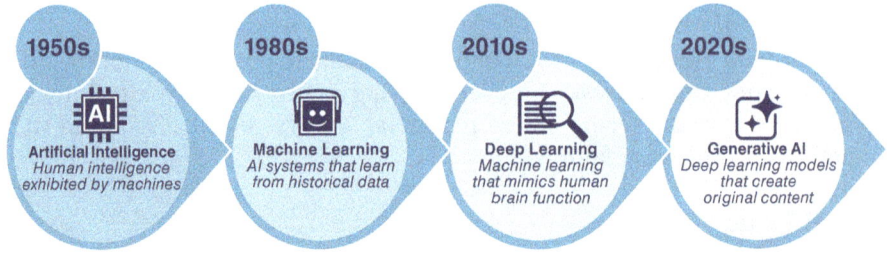

Source: Stryker and Kavlakoglu 2024

Figure 6-3   The evolution of machine learning

However, although the technology continues to change, one key element remains true for water and wastewater systems: Our physical infrastructure (pipes, pumps, tanks, motors, etc.) is the "crown jewel" that we must protect from cyber sabotage, regardless of the source. The application of CIE to today's (and future) automation systems helps asset owners protect these systems by design.

OUR CELESTIAL NAVIGATION

In February 2016, National Public Radio reported that the US Naval Academy was once again teaching office candidates celestial navigation. (Brumfiel, 2016). This was done because of an increasing recognition that the global positioning systems that modern naval vessels rely on are vulnerable. The capabilities had been phased out around a decade prior and thus created a huge vulnerability. If you can't navigate a naval vessel, how can it carry out its critical functions, whether it is fighting a war, delivering supplies, or tending to injured people after a natural disaster?

In many ways, this is an excellent analog for the overreliance on automation that has emerged for many utilities. The authors have had many experiences discussing A Day Without SCADA. We have experienced a wide range of responses from one utility in the San Francisco Bay Area who responded that their "control system has never worked right so, we pretty much call that Tuesday" to an IT manager in Oregon leaning back in her chair and cringing.

As the article asks, "So, why return now to the old ways?" We believe we have asked and answered that same question previously. For so many good reasons when it comes to building and maintaining resilient organizations, reducing unnecessary dependencies that allow for an improved capacity for operational disruptions of any kind is always a good thing. There is always the added benefit that the operations staff have a greater understanding of the system they are operating. Training opportunities abound.

CIE, for many organizations, will be merging the best of the old with the best of the new. For example, it provides the capabilities to operate manual and protective relays coupled with the ability to monitor network traffic, ingest threat intelligence provided by federal intelligence partners, and make operational decisions based on it. It is an ever-evolving world, whether or not we evolve with it.

REFERENCES

ANSI (American National Standards Institute)/AWWA. 2021. J100. Risk and Resilience Management of Water and Wastewater Systems. Denver: AWWA.

AWWA. 2025. "Digital Twins." https://www.awwa.org/resource/digital-twins/#about (accessed April 15, 2025).

Brumfiel, Geoff. 2016. U.S. Navy Brings Back Navigation by the Stars for Officers. https://www.npr.org/2016/02/22/467210492/u-s-navy-brings-back-navigation-by-the-stars-for-officers (accessed April 15, 2025).

Ohrt, A., A. Jones, J. Smith, V.L. Wright, B.R. Lampe, and R.V. Stolworthy. 2024. *Integrating Cyber-Informed Engineering into Enterprise Risk Management.* Idaho Falls, Idaho: Idaho National Laboratory. https://www.osti.gov/biblio/2480935 (accessed April 14, 2025).

Stryker, C. and E. Kavlakoglu. 2024. *What Is Artificial Intelligence (AI)?* IBM. https://www.ibm.com/think/topics/artificial-intelligence (accessed Nov. 29, 2024).

Wikipedia Contributors. 2025. *Failure of Imagination.* Wikipedia. Wikimedia Foundation. https://en.wikipedia.org/wiki/Failure_of_imagination (accessed Nov. 26, 2024).

Wikipedia Contributors. 2025. *Tiger Team.* Wikipedia. Wikimedia Foundation. https://en.wikipedia.org/wiki/Tiger_team (accessed Nov. 26, 2024).

US Environmental Protection Agency (USEPA). 2025. "America's Water Infrastructure Act: Risk Assessments and Emergency Response Plans." AWIA Section 2013/SDWA Section 1433. https://www.epa.gov/waterresilience/awia-section-2013 (accessed April 15, 2025).

Appendix A

CCE Case Study: Waterville Water District

DISCLAIMERS

Disclaimer #1

These case studies are fictionalized accounts of real water and wastewater utilities implementing elements of the consequence-driven, cyber-informed engineering (CCE) methodology. Names, locations, events, utilities, regions, countries, and incidents are fictitious. Any resemblance to actual utilities or events is purely coincidental.

Disclaimer #2

Any references to specific equipment, vendors, or technologies in this study do not imply increased susceptibility to cyberattack over other brands or devices. The equipment in this study is "typical" equipment often found in the industry. As a work of fiction, some features were modified to support the narrative.

INTRODUCTION

In this case study, we will examine how cyber-enabled sabotage could occur within a fictional organization. As the case study unfolds, the consequence-driven, cyber-informed engineering (CCE) methodology will be applied to demonstrate its purpose and applicability in finding ways to prevent one of the organization's worst-case scenarios.

UTILITY BACKGROUND

The fictitious Waterville Water District provides an average 14 million gallons per day (mgd) of water to 120,000 people in the southwestern United

States. Seasonally, water demand ranges from 8 to 20 mgd, which is the maximum production rate for the plant. The District maintains water supply across eight pressure zones of varying sizes.

In addition to retail customers, the District also serves Waterville University, a local military installation, a regional hospital with a Level II Trauma Center (ATS 2025), and one large commercial center.

The District's sole source of water is a local river. There is an emergency water supply connection with a neighboring utility. However, that connection has not been exercised in more than 20 years. The condition of the valving is unknown, and potential water chemistry effects are estimated to be minimal but have not been fully characterized. Both utilities use chlorine for disinfection. It is estimated that the emergency supply connection could serve the pressure zone that the hospital resides in.

While most of the year is relatively mild in temperature, July and August regularly experience several three-to-five-day periods of high temperatures greater than 100°F.

WATERVILLE WATER DISTRICT SYSTEM

Treatment

The District's water treatment plant (WTP) uses conventional water treatment technologies to provide safe, high-quality drinking water. The original WTP was constructed in the 1950s and has been modified and upgraded over the years to meet the changing regulations and customer base. The last major upgrade was conducted in 2009–2010. The process train is shown in Figure A-1.

While not directly applicable on the basis of geography, utility leadership has begun adopting the Ten State Standards (Water Supply Committee of the Great Lakes–Upper Mississippi River Board of State and Provincial Public Health and Environmental Managers 2012). The initial adoption effort was aimed at the chemical storage, and each treatment chemical has 30 days of storage at the average demand level (14 mgd).

Source water is pumped from the river into the headworks of the WTP. From there, the water flows via gravity to immediately before the dual-media filtration, where a pump station located within the fence boosts the flow. Water flows through the filters and into the disinfection chamber and then into the two finished water reservoirs to achieve the required disinfection contact time (CT) before entering the distribution system.

Finished water is pumped via the finished water pump station (FWPS) from the two finished water reservoirs to achieve the required disinfection CT before entering the distribution system. Without these reservoirs, the District would not be able to achieve the required CT. Automatically actuated valves can be used to isolate one or both reservoirs, if needed. These valves are more than 15 ft below grade. Various large pipelines within the plant convey water between processes.

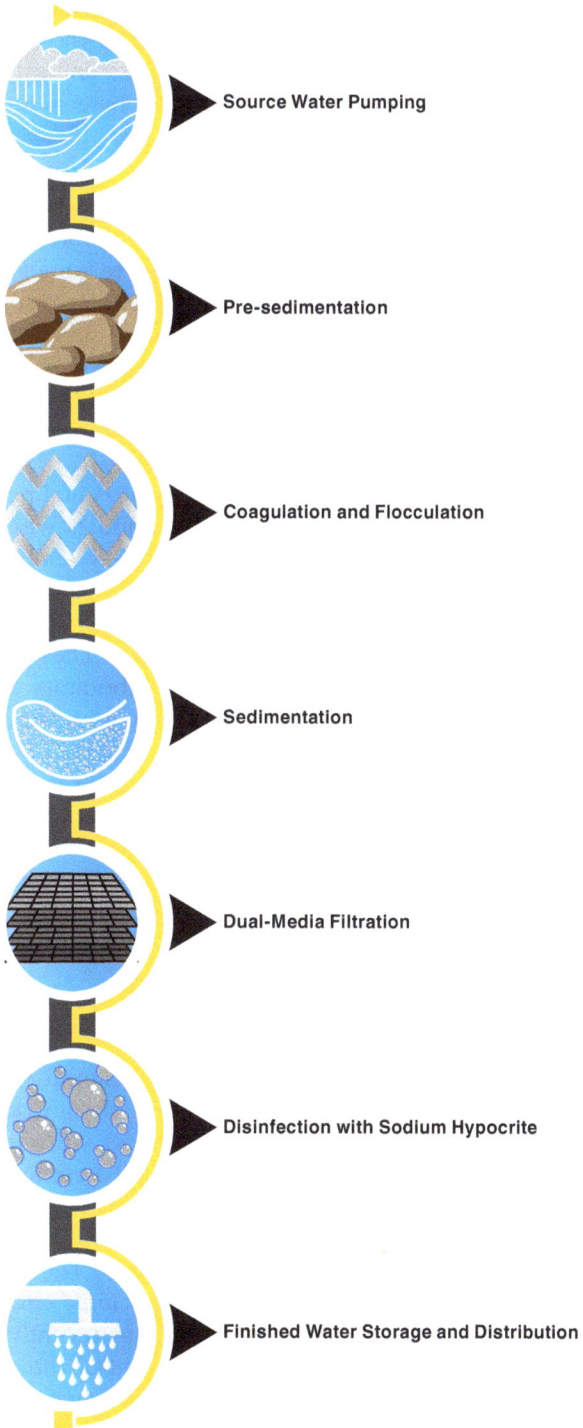

Source Water Pumping

Pre-sedimentation

Coagulation and Flocculation

Sedimentation

Dual-Media Filtration

Disinfection with Sodium Hypocrite

Finished Water Storage and Distribution

Figure A-1   Waterville Water District water treatment process

The WTP has a dedicated substation owned and operated by the District. An energy management system is connected to the supervisory control and data acquisition (SCADA) system to optimize energy usage through improved pump operations. The District has a standby generator at the plant that is due for replacement in the next five years.

Distribution

Finished water is distributed via the FWPS into the main pressure zone. There are two water towers within the in the main pressure zone for storage. Finished water is pumped from the main pressure zone up two series of booster pump stations and reservoirs to the highest-elevation pressure zones. There are two lower pressure zones to which water is passed from higher zones through pressure-reducing valves (PRVs). Both the military installation and hospital have dedicated pump stations to achieve the required distribution pressure for each customer.

Free chlorine concentrations, however, during seasonal periods of lower water usage may become an issue. The utility conducts weekly grab sampling at ~20 locations throughout the distribution system. This sampling regularly includes free chlorine and total coliform. On a monthly basis, the utility collects grab sampling for numerous other water quality parameters such as turbidity, various ions, and inorganic metals. These samples, along with most samples collected, are analyzed by the laboratory at the WTP.

It is estimated that at normal demand for July and August, the utility has ~12 hours of storage in the system.

The utility is contractually obligated to provide a set volume of water to both the hospital and military base. Because of regulatory and water rights hurdles, the hospital does not have a backup groundwater supply well. While the District has approached the hospital regarding participating in emergency preparedness exercises, this has not happened yet. In addition, the hospital will collect monthly grab samples of influent water for contract management purposes.

Control System

The District uses a third-party integrator, Electric Water, for all control system upgrades, programming support, and SCADA design document management. The integrator also provides 24/7 support via remote access to an engineering workstation on the plant network. Access to programs and design documents is available to customers on a supported server. Both the WTP and distribution system are controlled from the control room at the plant using a commercial off-the-shelf SCADA platform. The plant operations are controlled using Modicon M340 programmable logic circuits (PLCs). In an effort to reduce demands on staff, the integrator recently implemented a remote access solution allowing operators to control the plant remotely at night when the plant is unmanned. Alarm management continues to be an issue for operators at the plant. The District has implemented a policy requiring field verification for significant alarms. Minor alarms are generally addressed as a batch when operations staff are available.

Power System

The dedicated WTP substation is equipped with auxiliary direct current (DC) power system. The DC system comprises a battery management system (controller, alternating current [AC]/DC rectifier electronics, onboard maintenance bypass, and transfer capabilities), DC power distribution infrastructure (e.g., breakers, panels, wiring), a battery bank, and a resistive load bank. The DC system is a redundant system with multiple taps used to provide power to all substation control and protective devices, communications infrastructure, breakers, and switch actuators. If the DC power system is incapacitated (e.g., battery failure, controller failure, loss of AC power supply and charging), the ability to automatically and remotely control and monitor the substation is lost. The battery management system provides control and monitoring of the DC system, a configuration interface, communications capabilities, and battery bank charging functions. Battery health/charge is critical—from a degraded charge state, it can take up to 24 hours to restore batteries to a usable voltage level. The generator and power systems use Schneider Electric Modicon PLCs.

Additional Information

After a period of underinvestment in utility infrastructure in the 1980s and 1990s, utility staff are still working on rehabilitation and replacement of assets. From 2004 to 2007, the utility experienced several boil water notices in different parts of the service area, which caused a significant amount of turnover at the top of the organization. Customers, local politicians, and regulators recognize the utility's improved operations and proactive investment. However, between a series of rate increases to fund deferred infrastructure improvements and the reputational damage inflicted by prior management, they are still wary of any disruptions to their service. Current management continues to work hard to improve the District's reputation. Because of the progress in improving operations, data centers have been looking at locations within the service area. Financially, these would be a boon for the District. However, service disruptions would severely affect the District's reputation and may prevent the data centers from being constructed within the service area.

Public Information Disclosure

As part of a customer outreach strategy, the utility conducted a master planning exercise and posted most of the document on their website. The information posted includes maps of physical assets, proposed improvements to physical assets, and a staffing analysis. While they withheld the most sensitive information regarding their physical security and cybersecurity, to align with the procurement public disclosure requirements, they did note the SCADA software in use and the types of hardware, including PLCs and firmware requirements, in a recent request for bid. The District regularly hosts school and community groups for tours of the WTP.

CCE STEPS: WATERVILLE WATER DISTRICT CASE STUDY

Phase 1: Consequence Prioritization

Objective

Cause a public health/safety incident by distributing water with insufficient disinfection for more than 12 hours.

Scope

Meet the objective by focusing on the disinfection process.

Boundary Conditions

Cause a public health/safety incident by distributing unsafe water with insufficient disinfection for more than 12 hours.

Table A-1 summarizes the events identified. It also includes a brief description to describe the nature of the event.

Table A-1   Events and descriptions

Event	Event Description
No chlorination	The finished water is not chlorinated before being distributed to customers. This results in unsafe water throughout the distribution system.
Underchlorination	The finished water is underchlorinated and distributed to customers. This results in unsafe water in the far reaches of the distribution system.
Overchlorination	Excessive chlorine is added to the finished water and distributed to customers. This results in customer complaints and excessive use of chemical supplies.
Insufficient contact time	The finished water is chlorinated as normal but does not have sufficient CT. This results in unsafe water in the areas near the treatment plant.
Source water pump station inoperable	Source water cannot be pumped into the WTP for treatment. This results in no water being treated at the plant.
Finished water pump station inoperable	Finished water cannot be pumped into the distribution system for customer use. This results in no water being distributed to customers beyond what is in storage in the distribution system.
Hospital booster pump station inoperable	Finished water flow and pressure cannot be boosted to meet the needs of the hospital. Over time, this results in reduced medical care capabilities at that facility and affects patient care.
Military installation booster pump station inoperable	Finished water flow and pressure cannot be boosted to meet the needs of the military installation. Over time, this results in low flow and pressure for residential facilities at the installation.
Substation inoperable	The electrical substation is rendered inoperable, resulting in reliance on the backup power generation capabilities. This results in lower production that cannot meet demand and water restrictions.
Distribution system pressure transient	Pressure transients in the distribution system cause main damage. This results in localized property damage and service outages for customers.
Energy management system integrity failure	Pump operations are no longer optimized to minimize energy-related expenditures.

Table A-2 summarizes the cyber-events identified. It also includes a brief description to describe the nature of each cyber-event.

Table A-2   Cyber-events

Cyber-event	Cyber-event Description
1. No chlorination	Adversary gains access to the District control system environment and targets the disinfection process. Malicious modifications focus on the disinfectant dosage control. The controller logic is changed, so no chlorine is introduced to the treatment train. The attacker ensures that no indications (HMI) or notifications (alarming) are presented to the system operators.
2. Underchlorination	Adversary gains access to the District control system environment and targets the disinfection process. Malicious modifications focus on the disinfectant dosage control. The controller logic is changed, so insufficient chlorine is introduced to the treatment train. The attacker ensures that no indications (HMI) or notifications (alarming) are presented to the system operators.
3. Overchlorination	Adversary gains access to the District control system environment and targets the disinfection process. Malicious modifications focus on the disinfectant dosage control. The controller logic is changed, so excessive chlorine is introduced to the treatment train. The attacker ensures that no indications (HMI) or notifications (alarming) are presented to the system operators.
4. Insufficient contact time	Adversary gains access to the District control system environment and targets the disinfection process. Malicious modifications focus on the automatically actuated valves used to isolate the finished water reservoirs. The controller logic is changed, so the valves are closed when they should be open. The attacker ensures that no indications (HMI) or notifications (alarming) are presented to the system operators.
5. Source water pump station inoperable	Adversary gains access to the District control system environment and targets the source water pump station pumps. Malicious modifications focus on the source water pumps. The controller logic is changed, causing the pumps to operate in a manner that causes irreparable damage. The attacker ensures that no indications (HMI) or notifications (alarming) are presented to the system operators.
6. Finished water pump station inoperable	Adversary gains access to the District control system environment and targets the FWPS. Malicious modifications focus on the source water pumps. The controller logic is changed, causing the pumps to operate in a manner that causes irreparable damage. The attacker ensures that no indications (HMI) or notifications (alarming) are presented to the system operators.
7. Distribution system pressure transient	Adversary gains access to the District control system environment and targets the distribution system automatically actuated valves. Malicious modifications focus on the valves downgradient of the FWPS. The controller logic is changes causing the valves to close and open to quickly that causes damage to several water mains. The attacker ensures that no indications (HMI) or notifications (alarming) are presented to the system operators.

High-consequence event (HCE) severity scoring. Selected cyber-events are scored based on the criteria and weighting defined by the assessment team.

List of potential criteria:
- Effect on public safety
- Financial loss
- Reputational damage
- Effect on critical customers

Criteria weighting.
- Effect on public safety   High—3
- Financial loss   Medium—2
- Reputational damage   High—3
- Effect on critical customers   Medium—2

Combining the list of criteria with their assigned weighting values, the District develops and agrees on the following cyber-event scoring matrix shown in Table A-3.

Table A-3   Cyber-event severity scoring scheme

Criteria	Severity Scoring			
	None (0)	Low (1)	Medium (3)	High (5)
Effect on public safety $a = 3$	No effect on public safety	There is a low but definite risk to public safety—a few real or perceived illnesses occur.	There are a widespread number of illnesses but no deaths.	Widespread number of illnesses and one or more deaths
Financial loss $\beta = 2$	No financial loss	Financial losses are recoverable by the end of the next fiscal year.	Financial losses are substantial, they take several years to recover, and they displace other critical capital projects or result in rate increases.	Financial losses are substantial and require restructuring of the finances of the utility, including bonding, bankruptcy, and/or state/federal takeover.
Reputational damage $\delta = 3$	No reputational damage	Reputation is mildly damaged but recoverable by public relations investment over the next year.	Reputation is moderately damaged, new businesses within the service are delayed, and some management turnover is required.	Reputation is severely damaged, new businesses cancel plans to move into the area, management is fired, and state/federal oversight is implemented.
Effects on critical customers $\varepsilon = 2$	No effects on critical customers	Critical customer service is affected for <12 hours.	Critical customer service is affected for ~36 hours.	Critical customer service is affected for >72 hours.

Table A-4 summarized the cyber-event scoring for the no chlorination scenario.

Table A-4   Cyber-event scoring: No chlorination

Criteria	None (0)	Low (1)	Medium (3)	High (5)
Effect on public safety $a = 3$	No effect on public safety	There is a low but definite risk to public safety—a few real or perceived illnesses occur.	There are a widespread number of illnesses but no deaths.	Widespread number of illnesses and one or more deaths
Financial loss $\beta = 2$	No financial loss	Financial losses are recoverable by the end of the next fiscal year.	Financial losses are substantial, they take several years to recover, and they displace other critical capital improvement projects or result in rate increases.	Financial losses are substantial and require restructuring of the finances of the utility, including bonding, bankruptcy, and/or state/federal takeover.
Reputational damage $\delta = 3$	No reputational damage	Reputation is mildly damaged but recoverable by public relations investment over the next year.	Reputation is moderately damaged, new businesses within the service are delayed, and some management turnover is required.	Reputation is severely damaged, new businesses cancel plans to move into the area, management is fired, and state/federal oversight is implemented.
Effects on critical customers $\varepsilon = 2$	No effects on critical customers	Critical customer service is affected for <12 hours.	Critical customer service is affected for ~36 hours.	Critical customer service is affected for >72 hours.

Scoring Formula

$$Scored\ Impact\ Points$$
$$= \alpha(\text{Effect on Public Safety}) + \beta(\text{Financial Loss}) + \delta(\text{Reputational Damager})$$
$$+ \varepsilon(\text{Effect on Critical Customers})$$

$$Maximum\ Impact\ Points$$
$$= \alpha(5) + \beta(5) + \delta(5) + \varepsilon(5)$$
$$Maximum\ Impact\ Points = 3(5) + 2(5) + 3(5) + 2(5) = 50$$
$$HCE\ Severity\ Score$$
$$= \left(\frac{Scored\ Impact\ Points}{Maximum\ Impact\ Points} \right) \times 100$$

Event Scoring

Maximum points for HCE = 50

Event 1 No Chlorination:	$40/50 \times 100 = 80$
Event 2 Under Chlorination:	$10/50 \times 100 = 20$
Event 3 Excess Chlorination:	$12/50 \times 100 = 24$
Event 4 Insufficient CT:	$10/50 \times 100 = 20$

Table A-5 summarized the cyber-event scoring for the underchlorination scenario.

Table A-5   Cyber-event scoring: Underchlorination

Criteria	None (0)	Low (1)	Medium (3)	High (5)
Effect on public safety $a = 3$	No effect on public safety	There is a low but definite risk to public safety—a few real or perceived illnesses occur.	There are a widespread number of illnesses but no deaths.	Widespread number of illnesses and one or more deaths
Financial loss $\beta = 2$	No financial loss	Financial losses are recoverable by the end of the next fiscal year.	Financial losses are substantial, they take several years to recover, and they displace other critical capital improvement projects or result in rate increases.	Financial losses are substantial and require restructuring of the finances of the utility, including bonding, bankruptcy, and/or state/federal takeover.
Reputational damage $\delta = 3$	No reputational damage	Reputation is mildly damaged but recoverable by public relations investment over the next year.	Reputation is moderately damaged, new businesses within the service are delayed, and some management turnover is required.	Reputation is severely damaged, new businesses cancel plans to move into the area, management is fired, and state/federal oversight is implemented.
Effects on critical customers $\varepsilon = 2$	No effects on critical customers	Critical customer service is affected for <12 hours.	Critical customer service is affected for ~36 hours.	Critical customer service is affected for >72 hours.

Table A-6 summarized the cyber-event scoring for the no chlorination scenario.

Table A-6   Cyber-event scoring: Excess chlorination

Criteria	Severity Scoring			
	None (0)	Low (1)	Medium (3)	High (5)
Effect on public safety $a = \underline{3}$	No effect on public safety	There is a low but definite risk to public safety—a few real or perceived illnesses occur.	There are a widespread number of illnesses but no deaths.	Widespread number of illnesses and one or more deaths
Financial loss $\beta = 2$	No financial loss	Financial losses are recoverable by the end of the next fiscal year.	Financial losses are substantial, they take several years to recover, and they displace other critical capital improvement projects or result in rate increases.	Financial losses are substantial and require restructuring of the finances of the utility, including bonding, bankruptcy, and/ or state/federal takeover.
Reputational damage $\delta = \underline{3}$	No reputational damage	Reputation is mildly damaged but recoverable by public relations investment over the next year.	Reputation is moderately damaged, new businesses within the service are delayed, and some management turnover is required.	Reputation is severely damaged, new businesses cancel plans to move into the area, management is fired, and state/ federal oversight is implemented.
Effects on critical customers $\varepsilon = \underline{2}$	No effects on critical customers	Critical customer service is affected for <12 hours.	Critical customer service is affected for ~36 hours.	Critical customer service is affected for >72 hours.

Table A-7 summarized the cyber-event scoring for the insufficient CT scenario.

Table A-7   Cyber-event scoring: Insufficient CT

Criteria	Severity Scoring			
	None (0)	Low (1)	Medium (3)	High (5)
Effect on public safety $a = 3$	No effect on public safety	There is a low but definite risk to public safety—a few real or perceived illnesses occur.	There are a widespread number of illnesses but no deaths.	Widespread number of illnesses and one or more deaths
Financial loss $\beta = 2$	No financial loss	Financial losses are recoverable by the end of the next fiscal year.	Financial losses are substantial, they take several years to recover, and they displace other critical capital improvement projects or result in rate increases.	Financial losses are substantial and require restructuring of the finances of the utility, including bonding, bankruptcy, and/or state/federal takeover.
Reputational damage $\delta = 3$	No reputational damage	Reputation is mildly damaged but recoverable by public relations investment over the next year.	Reputation is moderately damaged, new businesses within the service are delayed, and some management turnover is required.	Reputation is severely damaged, new businesses cancel plans to move into the area, management is fired, and state/federal oversight is implemented.
Effects on critical customers $\varepsilon = 2$	No effects on critical customers	Critical customer service is affected for <12 hours.	Critical customer service is affected for ~36 hours.	Critical customer service is affected for >72 hours.

Table A-8 summarizes the event scoring completed for each cyber-event.

Table A-8   High-consequence event scoring summary.

Event	Description	Impact Points	HCE Severity Score
1	No chlorination	40	80
2	Underchlorination	10	20
3	Over chlorination	12	24
4	Insufficient CT	10	20

HCE Identification

Event: No chlorination. Adversary gains access to the District control system environment and targets the disinfection process. Malicious modifications focus on the disinfectant dosage control. The controller logic is changed, so no chlorine is introduced to the treatment train. The attacker ensures that no indications (human–machine interface [HMI]) or notifications (alarming) are presented to the system operators.

Phase 2: System-of-Systems Analysis

Creating an HCE Block Diagram

The starting point for phase 2, System-of-Systems Analysis (SoS Analysis), is the creation of a relatively simple, high-level block diagram for each HCE to help with visualizing the cyber manipulation required to accomplish the outcome. This exercise helps narrow the scope of analysis, organizes the physical and functional connections between the target components and the affected systems, and minimizes the volume of information collected to describe each HCE. The block diagram provides a starting point for identifying what information and system accesses the adversary needs to accomplish the HCE and will be used to define and organize the data collection efforts. Block diagrams are provided for each HCE.

HCE 1: No Chlorination (Description)

Adversary gains access to the District control system environment and targets the disinfection process. Malicious modifications focus on the disinfectant dosage control. The controller logic is changed, so no chlorine is introduced to the treatment train. The attacker ensures that no indications (HMI) or notifications (alarming) are presented to the system operators. The block diagram for this HCE is shown in Figure A-2.

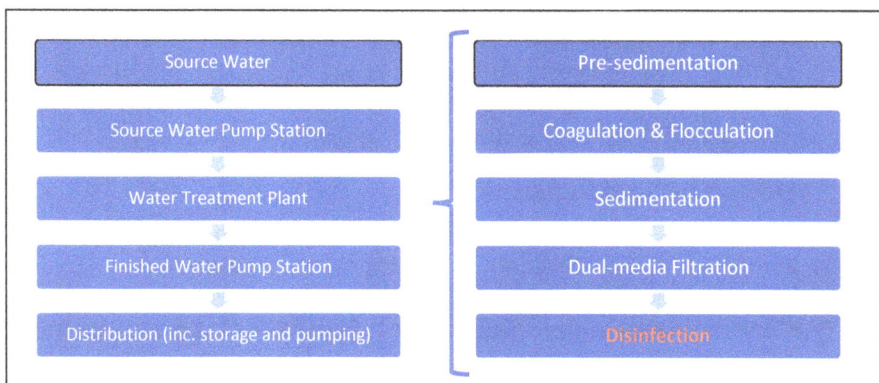

Figure A-2   HCE 1: No chlorination (HCE block diagram)

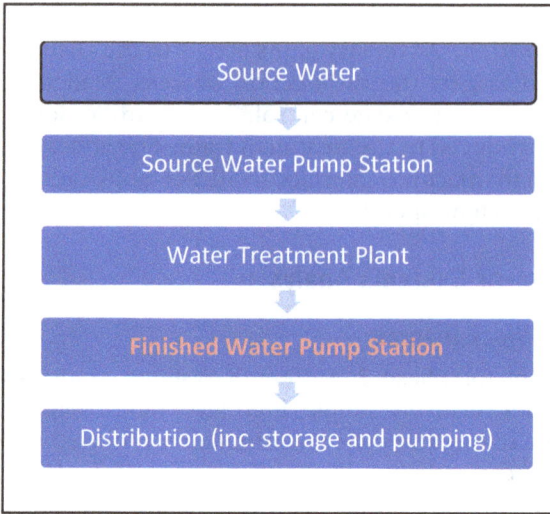

Figure A-3   HCE 2: FWPS (HCE block diagram)

HCE 2: FWPS Damage (Description)

Adversary gains access to SCADA and targets the FWPS controller, excessively cycling and damaging enough pumps to reduce capacity. Alternatively, the FWPS is operated to run-dry condition and burn out the pumps. The block diagram for this HCE is shown in Figure A-3.

HCE 3: Pressure Transients (Description)

Adversary hacks into the third-party integrator during the night because no operators monitor the system. The adversary gains access to the engineering workstation on the plant network. The adversary configures the valve to remain closed and covers alarming. The adversary also hides the status of the physical valve position by displaying it as normal to operations. Using the third party, the adversary has access to all Waterville operations, and that unverified trust is taken advantage of so Waterville doesn't suspect anything. Adversary knows the SCADA because it is off-the-shelf and commercial. The block diagram for this HCE is shown in Figure A-4.

"Perfect Knowledge" and Its Benefits

Phase 2 continues with a deeper look at the production or business function(s) identified in the HCE block diagrams, presented previously. The entity (in this case, Waterville) knows best how it delivers these functions, and CCE leverages that unique and in-depth understanding from a technical and operational perspective.

Waterville will focus on identifying, assessing, and categorizing "artifacts" that describe details of the systems, system configurations, system operations, supply chain, and other personnel support activities present in

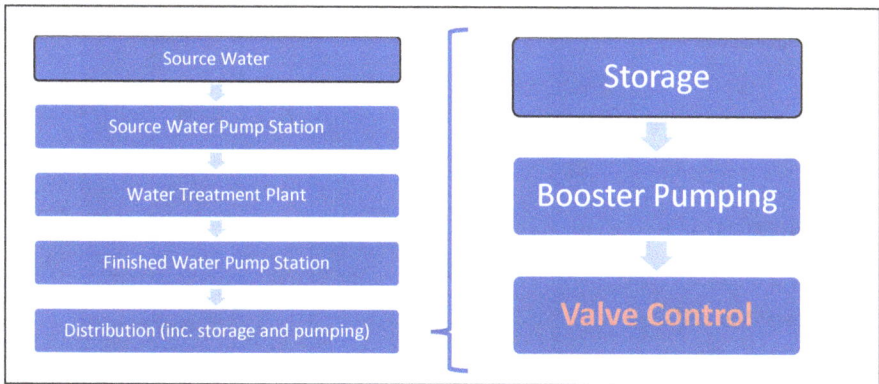

Figure A-4 HCE 3: Pressure transients (HCE block diagram)

delivery of the pasteurization function that is targeted in the HCE. Waterville must iteratively consider the following:

- What systems and components are involved in the HCE?
- What documentation is needed to describe interconnected systems and dependencies?
- What relationships with other entities are involved?

Building an SoS analysis starts with the system/component that must be affected to cause the HCE and works outward. Success here is measured by developing "perfect knowledge" of the system(s), operations, and support. Artifacts collected should include details of the target system(s), including logical and physical connectivity, system dependencies, controllers, technical and operational manuals, engineering/process/communications diagrams, protocols, access lists, associated manufacturers, trusted relationships, contractors, suppliers, emergency procedures, and personnel lists, among others.

Keep in mind the objective is to identify not only the technical and operational details but also where the information is documented and stored. The project team must determine if the critical information is stored only on internal servers or is also available on public-facing assets.

Establishing "perfect knowledge" is most accurately and efficiently executed through the development and use of a functional taxonomy.

Functional Taxonomy

Background/purpose. The CCE functional taxonomy is a relational framework used for describing an organization's critical functions; the people, process, and technology (PPT) that enable those critical functions; and the artifacts that document an entity's unique implementation for function delivery. Most importantly, the taxonomy maps the organization's critical functions and those enabling functions that support them. The artifacts and their relative location within the taxonomy framework describe the "what," "where," "when," "how," and "who" of the HCE.

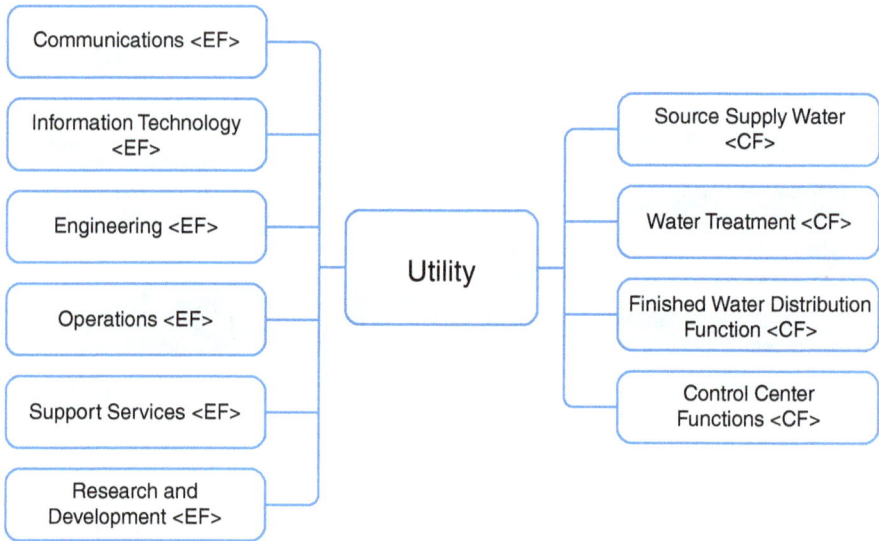

CF—critical function, EF—enabling function

Figure A-5   Top-level CFs and EFs

Method. Specialized visualization applications like MindManager* are effective tools for capturing the hierarchical organization and mapping of relationships between objects. If specialized applications are not available to the CCE Team, taxonomy mapping can also be accomplished using a spreadsheet like Excel.

Approach. CCE taxonomy mapping begins by naming the entity (Waterville) at the root level of the taxonomy structure and creating the entity box (see Figure A-5). Every object created in the mapping from here forward will describe a part or function of Waterville (e.g., business, business subsector).

From here, a CCE taxonomy clearly divides the functional mapping into two halves. The right-hand side is branches of critical functions that describe the actions or activities making up Waterville's primary purpose. If any of these critical functions are disrupted by an adversary, it would result in the worst of "bad days" for Waterville.

The left-hand side is reserved for mapping enabling function branches. Enabling functions describe the infrastructure, PPT, and systems Waterville uses to both physically and logically deliver its critical functions.

A functional taxonomy mapping for Waterville is shown in Figures A-6 through A-8. These figures demonstrate how Waterville has located elements and created branches required to document its critical functions and the delivery of these functions.

---

* Visit www.mindmanager.com/en/product/mindmanager/ to learn more about MindManager's mind mapping software.

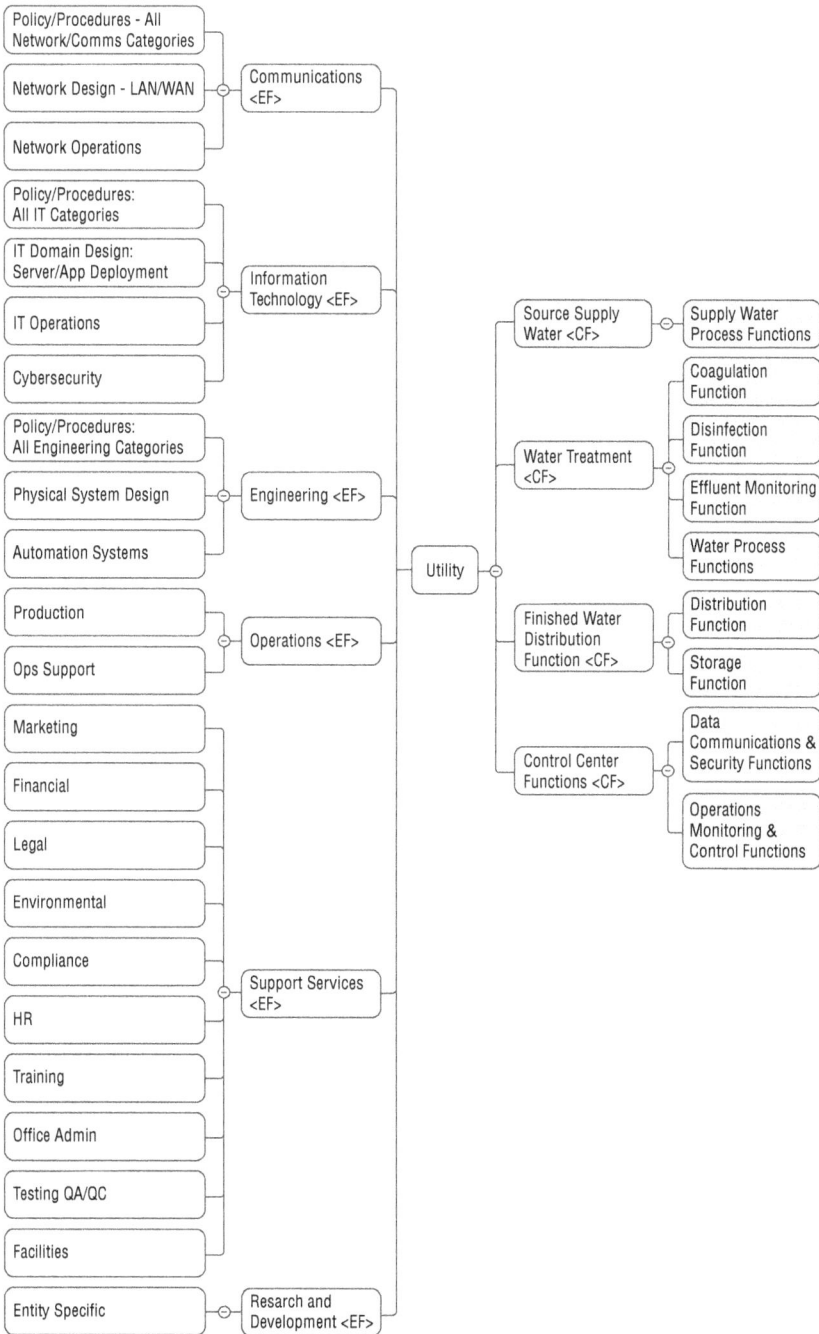

CF—critical function, EF—enabling function, HR—human resources, IT—information technology, QA—quality assurance, QC—quality control

Figure A-6 **Expansion of Waterville's functional taxonomy with more developed CFs (right) and EFs (left)**

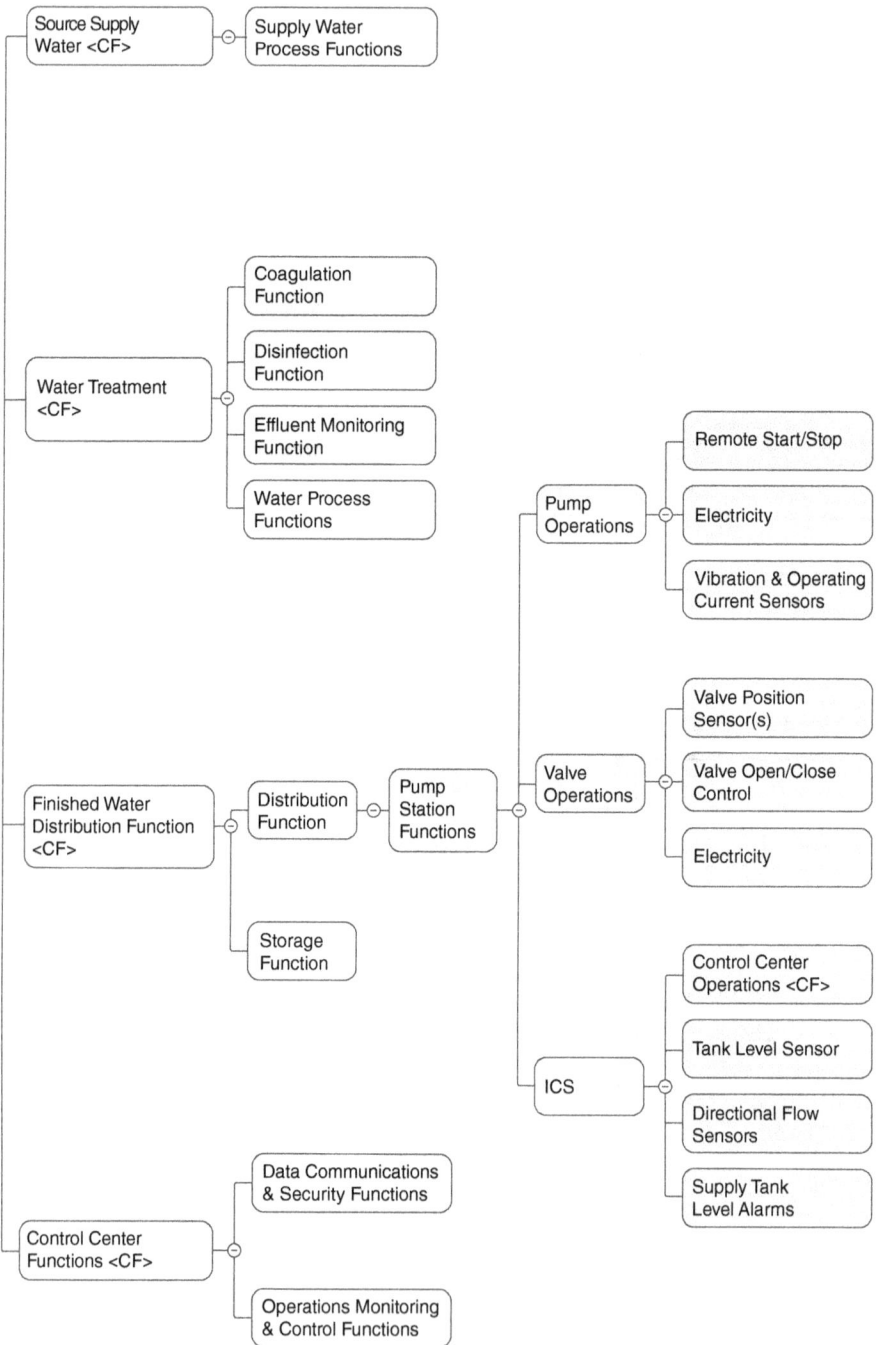

CF—critical function, ICS—industrial control systems

Figure A-7   Further expansion of the right-hand side (CFs) for Waterville's functional taxonomy related to disinfection

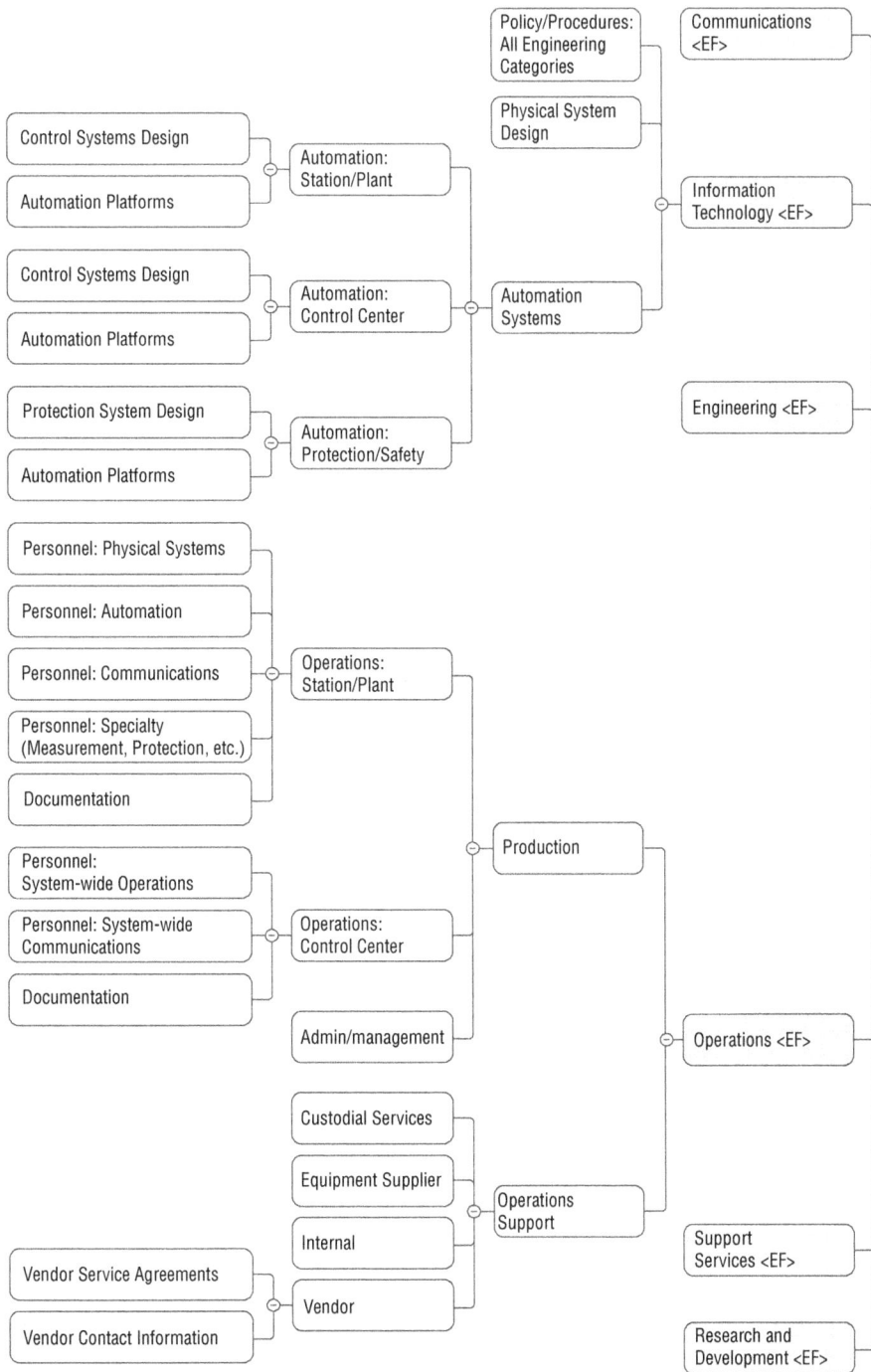

Figure A-8 Further expansion of the left-hand side (enabling functions [EFs]) for Waterville's functional taxonomy related to the control system

As a general note, the language used may not be universal across the water sector, and practitioners (e.g., engineers, technicians, and operators) from specific entities may use differing terms.

System Description

A system description is a summary of all the information Waterville gathered in phase 2. This description should summarize the functional taxonomy mapping and provide traceability to where all the information resides—as well as who has access to it. This will be the output of phase 2 and the input to phase 3. A system description for this use case is detailed in the following sections.

System description: Water production. The WTP uses conventional water treatment technologies to produce drinking water. The original WTP was constructed in the 1950s and has been modified and upgraded over the years to meet the changing regulations and customer base. The last major upgrade was conducted in 2009–2010. The process train is shown in Figure A-1.

The utility has adopted the Ten State Standards for chemical storage, resulting in 30 days of storage at average demand (14 mgd).

Source water is pumped from the river into the headworks of the WTP. From there, the water flows via gravity to immediately before the dual-media filtration, where a pump station located within the fence boosts the flow. Water then flows through the filters and into the disinfection chamber and then into the two finished water reservoirs to achieve the required disinfection CT before entering the distribution system.

Finished water is pumped via the FWPS from the two finished water reservoirs to achieve the required disinfection CT before entering the distribution system. Without these reservoirs, the District would not be able to achieve the required CT. Automatically actuated valves can be used to isolate one or both reservoirs, if needed. These valves are more than 15 ft below grade. Various large pipelines within the plant convey water between processes.

The WTP has a dedicated substation owned and operated by the District. An energy management system is connected to the SCADA system to optimize energy usage through improved pump operations. The District has a standby generator at the plant that is due for replacement in the next five years.

The HCE block diagram in Figure A-9 shows additional production details and sequences for the treatment and distribution of water.

System description: SCADA. The SCADA control room is located at the WTP. Both treatment and distribution are controlled from this location. Production and distribution are primarily controlled by throttling pumping throughout the plant. Operators are responsible for changing any pumping.

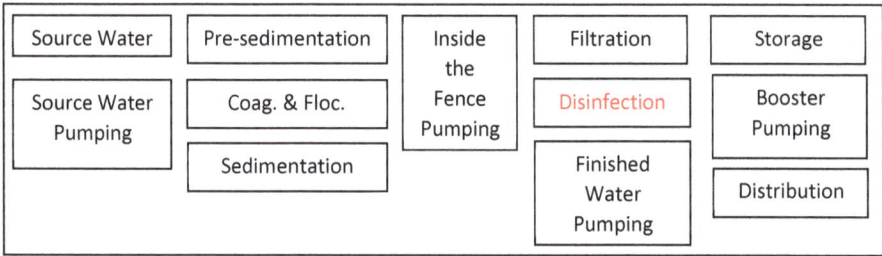

Figure A-9 A further-developed block diagram showing water treatment details and sequence for the entire water production and distribution process

The SCADA network is separated from the enterprise IT network by a single Palo Alto firewall. Within the WTP, the District has standardized to the Modicon M340 PAC PLC. Allen-Bradley PanelView Plus 6 Graphic Terminals are used for the HMI panel display. These HMIs require a development project that runs on the HMI. Cradlepoint devices (Ericsson Enterprise Solutions) are used to communicate with remote facilities which have a variety of small PLCs.

The SCADA network is composed of segmented loops aligned to the critical pumping and processes within the plant. The distribution system is a hub-and-spoke model network.

The District's America's Water Infrastructure Act (AWIA) risk and resilience assessment noted that significant upgrades are required to meet current best practices. These include implementing encryption standards on wired and nonwired communications, improving the network architecture, training on cybersecurity awareness, maintaining up-to-date and supported software versions (e.g., PLC firmware, SCADA software, OS patches, etc.), and developing policies and procedures.

System description: SCADA software platform. The District uses Ignition by Inductive Automation as the SCADA software to enable all automation and remote monitoring and control of physical assets in the system. The Ignition architecture consists of a pair of redundant Ignition Gateway servers that are located within the same server rack, with several Ignition clients installed on workstations and operator interface terminals (OITs) located throughout the treatment plant. The Ignition client on the workstations and OITs is dependent on the Ignition Gateway servers for monitoring and control.

To prevent reliance on a single entity for SCADA maintenance and support, the District often maintains support contracts with two separate vendors to update their SCADA software. One of these vendors provides primary support, and the second provides emergency support only. The vendors have the option of either performing SCADA updates remotely using their own computers and or updating on-site. When updating remotely to the SCADA server, a virtual private network (VPN) connection is used. When on-site, policy is to use a USB provided by the utility to transfer the files from the support computer to the SCADA server.

Once the updates are deployed to the servers, the Ignition Gateway pushes the updates to the Ignition client installed on all workstations and OITs, and the changes take place immediately. Once updates are completed, the vendors often store the latest SCADA files on their respective computers. These computers are used to provide SCADA programming and configuration services for various other clients throughout the country.

System description: Disinfection. Sodium hypochlorite (NaOCl) (and all other chemicals) is added on a continuous basis while drinking water is being produced. The disinfectant dosage is controlled by operators throttling pump stroke and percent speed to calculate the volumetric flow.

NaOCl is the second to last chemical introduced into the process stream before being pumped by the FWPS into the distribution system. The last chemical added is caustic soda for pH control.

NaOCl concentrations at the water entering the distribution system range from 1.3 to 2.0 mg/L. This concentration is monitored by an online water Hach CL17sc Colorimetric Chlorine Analyzer. The target free chlorine concentration range at the furthest points of the distribution system from the WTP is 0.3–0.4 mg/L. Because of poor flow regimes in some areas, occasional flushing is required to maintain the proper chlorine residual.

The operator interface is an Allen-Bradley PanelView Plus 6 Graphic Terminal. "Low Flow" and "Residual Only" modes are available to operators, if needed. Dosage may be controlled at the local panel but is overridden by changes from the main control room. The major control elements of the disinfection process system are shown on the HMI graphic (see Figure A-10) and the process and instrument diagram (see Figure A-11).

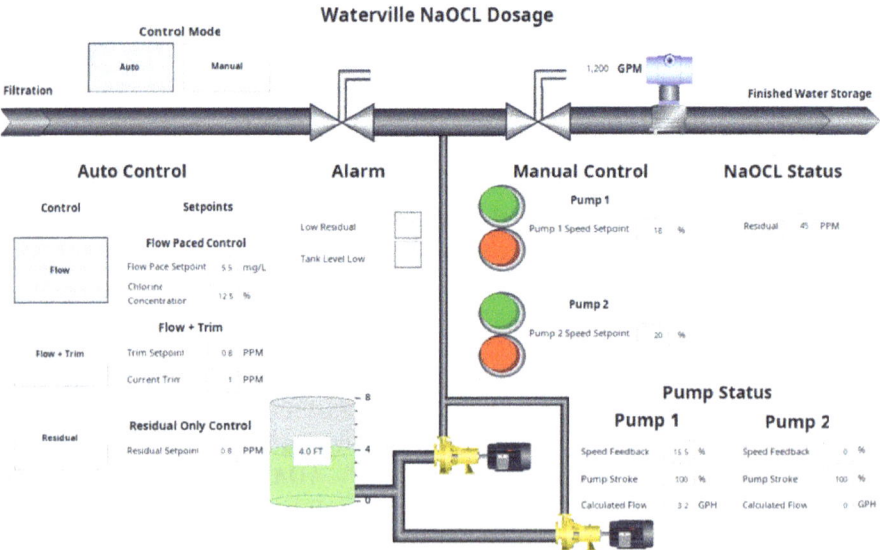

Figure A-10   NaOCl disinfection process HMI screen

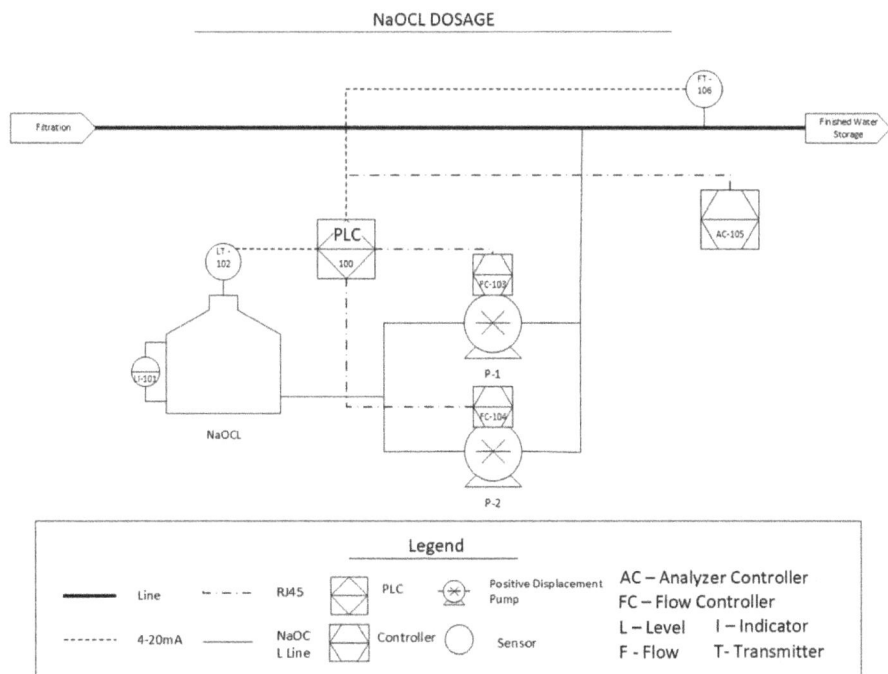

Figure A-11   NaOCl disinfection process and instrumentation diagram

System description: FWPS. Three 500-hp pumps are used to pump water from the WTP to the distribution system. Two of these pumps are required for water service during medium to high production. The third is a spare, "hot" standby on the same PLC, available for operators to quickly increase pumping or swap operational pumps, if needed. Each pump has a variable-frequency drive attached to it for energy savings and operational flexibility.

Pumping is manually controlled to meet demand and maintain sufficient storage in the distribution system. High-level alarms in the reservoirs may shut down pumping and alert the operator to respond. Low-level alarms alert the operator to respond.

Pumping may be controlled remotely from the control center, the remote access portal, or the local HMI panel. The major control elements of the FWPS are shown on the HMI graphic (see Figure A-12).

System description: Valve control. Currently, the District has an automatic valve at the FWPS, at two critical points in the distribution system, and at two of the District's reservoirs. The two reservoirs are currently equipped with seismic valves that can also be remotely opened and closed. The automatic valves are rarely used and are exercised approximately once every five years.

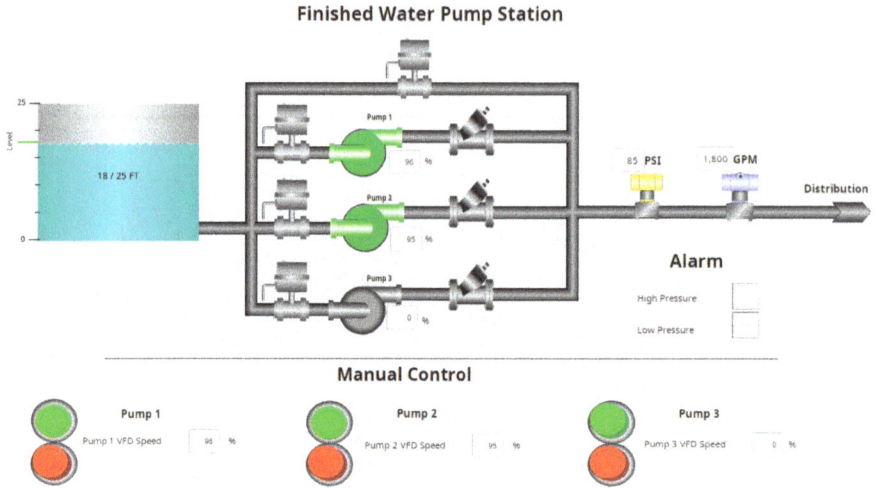

VFD—variable-frequency drive

Figure A-12 Finished water pump station HMI screen

System description: Operations. Water system operators are on-site at the WTP for a 10-hour shift, 6 a.m. to 4 p.m., Monday through Friday. When operators are not on-site, including Saturday and Sunday, the on-call operator monitors the system water remotely. During the weekday shift, system operators are responsible for implementing a set of operating procedures, routine maintenance, and operations tasks.

Tasks include the following:
- Daily:
  - Conduct three walks of the plant per shift. These start at the source water pumps, follow the process, and finish at the FWPS. These are done at the beginning of the shift, the middle of the shift, and the end of the shift.
  - Visually inspect all major assets, recording any changes in condition/status. These include pumps, tanks, valves, control panels, and electrical assets.
  - Check/record WTP HMI readings, including water chemistry parameters, flow rate, pump operational parameters, chemical usage and remaining volume, filter backflow status, and valve positions. Operators also confirm physical security measures are in place and functioning.
  - Monitor distribution systems parameters at main control room. Parameters include current system demand, reservoir water levels, valve positions, and system pressures.

- Weekly:
  - ° Conduct proactive maintenance on assets.
  - ° Calibrate sensors.
  - ° Manually collect free chlorine and total coliform grab samples from 20 locations within the distribution system. This is generally done on Tuesday of each week. Samples are analyzed in-house with validated and verified data reported by Wednesday afternoon.
  - ° Visit and visually inspect remote facilities including reservoirs, pump stations, and PRVs.
- As-needed/less frequent activities:
  - ° Conduct reactive maintenance on assets that require it.
  - ° Maintain the online chlorine analyzer (monthly).
  - ° Perform emergency response operations.

System description: External support. The control system integrator, Electric Water, provides 24/7 remote support for the District. They provide all integration and support services for the water system and heating, ventilating, and air-conditioning (HVAC) systems. The integrator stores all design, configuration, programming, and integration files for the District. These files are remotely accessible to the District via a supported SharePoint server portal.

The program developed by Electric Water also provides the remote access solution for operations staff to monitor and control the water system while not on-site. This program allows operators to monitor and control water flow, chemical dosing, alarms, trend analysis, and physical security camera feeds.

Network connectivity allowing vendor support to the District's water system is shown on the "Network architecture" graphic (see Figure A-13) and the "Network architecture—Integrator's access" graphic (see Figure A-14).

Phase 3: Consequence-Based Targeting

The system description developed in phase 2 will be the starting point for phase 3 as we examine the system from an adversarial perspective. We will determine what elements of the system the adversary needs to manipulate to achieve the HCE.

For any given HCE, there are likely many paths an adversary could take to achieve a particular outcome. Thus, a primary goal of phase 3 is identifying any points that an adversary must access or traverse—known as "choke points." Choke points are ideal locations to implement potential mitigations and protections because they effectively provide a way to cut off (or at least detect) the adversary's progress.

For the purposes of phase 3, we will consider two stages of adversary activity: payload development and payload deployment. For this HCE, and for each of these stages, we use phase 2's system description (and supporting documentation as needed) to help identify the adversary's

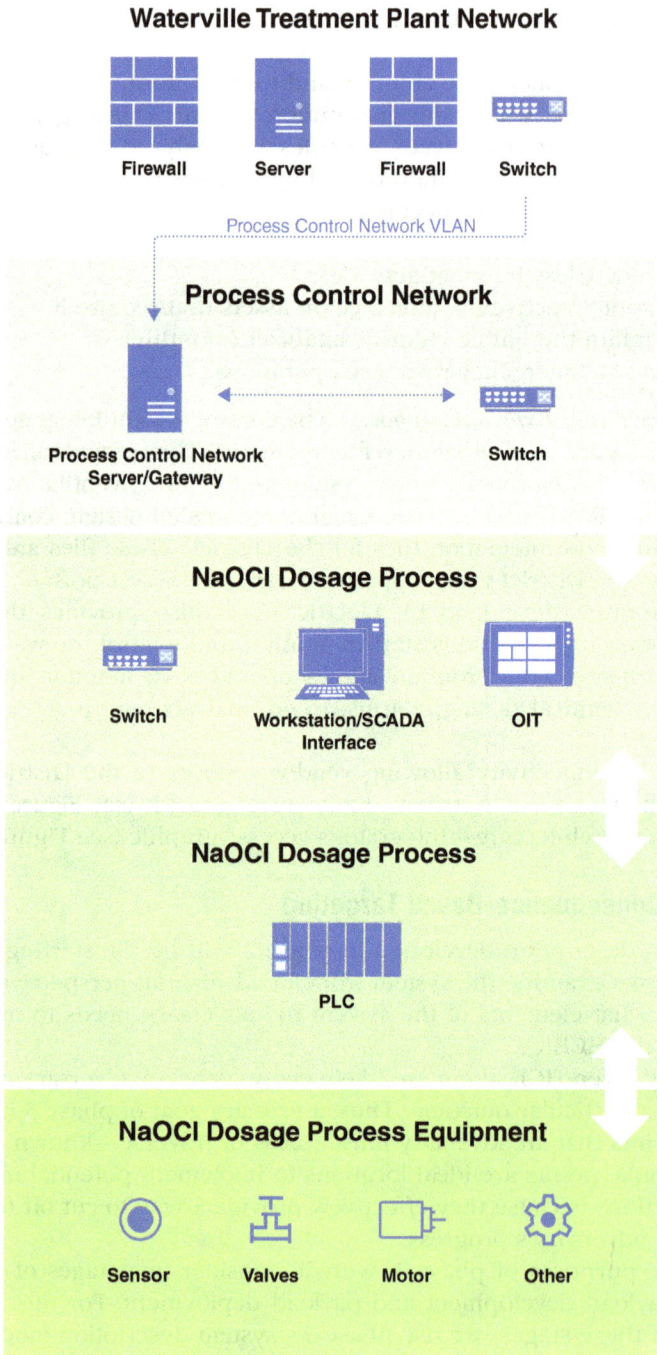

Waterville Treatment Plant Network

Firewall Server Firewall Switch

Process Control Network VLAN

Process Control Network

Process Control Network
Server/Gateway

Switch

NaOCl Dosage Process

Switch Workstation/SCADA Interface OIT

NaOCl Dosage Process

PLC

NaOCl Dosage Process Equipment

Sensor Valves Motor Other

OIT—operator interface terminal, PLC—programmable logic circuit, SCADA—supervisory control and data acquisition

Figure A-13 Network architecture

Waterville Treatment Plant Network

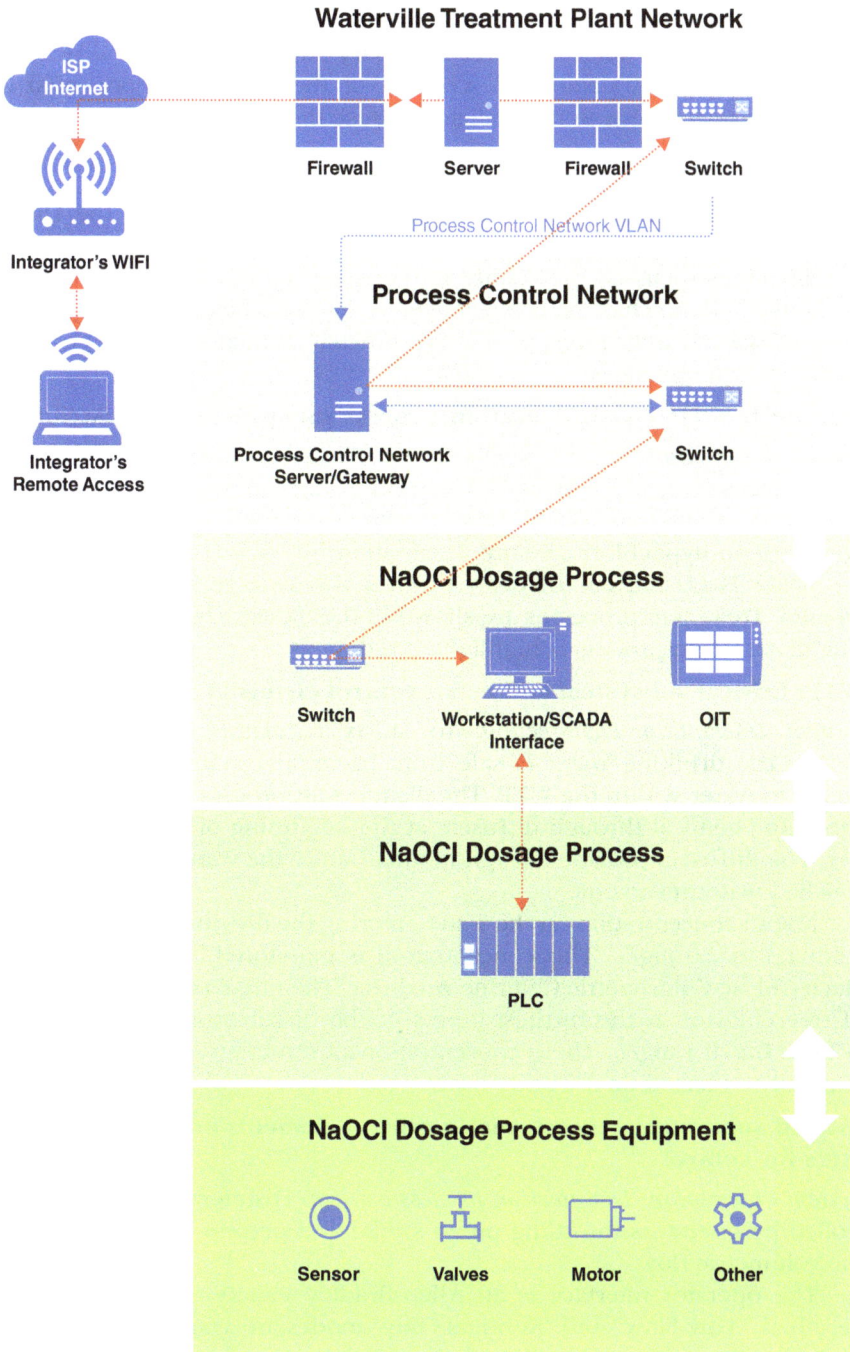

Firewall   Server   Firewall   Switch

Process Control Network VLAN

Process Control Network

Process Control Network
Server/Gateway

Switch

NaOCl Dosage Process

Switch   Workstation/SCADA
Interface   OIT

NaOCl Dosage Process

PLC

NaOCl Dosage Process Equipment

Sensor   Valves   Motor   Other

OIT—operator interface terminal, PLC—programmable logic circuit, SCADA—supervisory control and data acquisition

Figure A-14   Network architecture—Integrator's access

- *system targeting description*: combination of the system description and the system analysis for targeting;
- *technical approach*: what the adversary must do to cause the HCE; and
- *target details*: detailed description of the system component(s) the adversary needs to manipulate to cause the HCE.

Development Stage

System targeting description: HCE. Adversary gains access to the District control system environment and targets the disinfection process. Malicious modifications focus on the disinfectant dosage control. The controller logic is changed, so no chlorine is introduced to the treatment train. The attacker ensures that no indications (HMI) or notifications (alarming) are presented to the system operators.

Waterville WTP: Physical location, process subsystems, and distribution

System description: WTP. The District's WTP uses conventional water treatment processes to safely treat and distribute water to its customers. To meet regulatory requirements and ensure water safety, all water is disinfected with sodium hypochlorite before distribution.

Water is treated continuously to meet the average annual demand of 14 mgd. Treatment processes are shown in the block diagram of Figure A-2. The disinfection process is highlighted red.

WTP: Critical subsystem and major control elements

System description: Disinfection. To satisfy regulatory requirements and ensure the drinking water is safe from bacterial contamination NaOCl is added to water within the WTP. The disinfectant process is designed to add the bulk chemical through diffusers at the beginning of the mixing chamber. The diffusers provide enough mixing before the water flowing into the finished water reservoirs.

NaOCl concentrations at the water entering the distribution system range from 1.3 to 2.0 mg/L. This concentration is monitored by an online water Hach CL17sc Colorimetric Chlorine Analyzer. The target concentration range of free chlorine at the furthest points of the distribution system from the WTP is 0.3–0.4 mg/L. These concentrations are measured by grab samples manually collected on a weekly basis.

Critical subsystem process: major control elements and critical parameters for control

System description: Disinfection process control. Disinfectant dosing is controlled by operators throttling pump stroke and percent speed to calculate the volumetric flow.

The operator interface is an Allen-Bradley PanelView Plus 6 Graphic Terminal. "Low Flow" and "Residual Only" modes are available to operators, if needed. Dosage may be controlled at the local panel but is overridden by changes from the main control room.

Critical subsystem process control: Controller and HMI identification, process screenshot (HMI)

System description: Disinfection system controls platform. The District uses Ignition by Inductive Automation as the SCADA software to enable all automation and remote monitoring and control of physical assets in the system. This includes the disinfection process.

The Ignition architecture consists of a pair of redundant Ignition Gateway servers that are located within the same server rack, with several Ignition clients installed on workstations and OITs located throughout the treatment plant. The Ignition client on the workstations and OITs is dependent on the Ignition Gateway servers for monitoring and control.

The vendors have the option of either performing SCADA updates remotely using their own computers or updating on-site. When updating remotely to the SCADA server, a VPN connection is used. When on-site, the policy is to use a USB thumb drive provided by the utility to transfer the files from the support computer to the SCADA server.

Once the updates are deployed to the servers, the Ignition Gateway pushes the updates to the Ignition client installed on all workstations and OITs, and the changes take place immediately. Once updates are completed, the vendors often store the latest SCADA files on their respective computers. These computers are used to provide SCADA programming and configuration services for various other clients throughout the country.

Critical subsystem operations: Process parameter monitoring and process visibility source

System description: Operations. Water system operators are on-site at the WTP for a 10-hour shift, 6 a.m. to 4 p.m., Monday through Friday. When operators are not on-site, including Saturday and Sunday, the on-call operator monitors the system water remotely. During the weekday shift, system operators are responsible for implementing a set of operating procedures, routine maintenance, and operations tasks. Operations tasks relevant to the disinfection process include the following:

Daily:
1. Conduct three walks of the plant per shift. These start at the source water pumps, follow the process, and finish at the FWPS. These are done at the beginning of the shift, the middle of the shift, and the end of the shift.
2. Visually inspect all major assets, recording any changes in condition/status. These include pumps, tanks, valves, control panels, and electrical assets.
3. Check/record WTP HMI readings, including water chemistry parameters, flow rate, pump operational parameters, chemical usage and remaining volume, filter backflow status, and valve positions. Operators also confirm physical security measures are in place and functioning.

4. Monitor distribution systems parameters at main control room. Parameters include current system demand, reservoir water levels, valve positions, and system pressures.

Weekly:
1. Calibrate sensors.
2. Manually collect free chlorine and total coliform grab samples from 20 locations within the distribution system. This is generally done on Tuesday of each week. Samples are analyzed in-house with validated and verified data reported by Wednesday afternoon.
3. Visit and visually inspect remote facilities including reservoirs, pump stations, and PRVs.

Production automation: External technical support (Electric Water) and remote access capabilities

System description: External technical support for control system. The control system integrator, Electric Water, provides 24/7 remote support for the District. They provide all integration and support services for the water system and HVAC systems. The integrator stores all design, configuration, programming, and integration files for the District. These files are remotely accessible to the District via a supported SharePoint server portal.

The program developed by Electric Water also provides the remote access solution for operations staff to monitor and control the water system while not on-site. This program allows operators to monitor and control water flow, chemical dosing, alarms, trend analysis, and physical security camera feeds.

Network connectivity allowing vendor support to the District's water system is shown on the "Network architecture—Integrator's access" graphic (see Figure A-15).

System Analysis for Targeting

A system analysis for targeting is an additional analysis of key systems, components, people, processes, digital connectivity, and data flows that "fill in the gaps" and enable an adversary to assemble a relationally contiguous system targeting description for attack.

Key additional targeting information and steps (reconnaissance—open source and target environment)

Technical Support: Integrator

A press release indicates that Waterville's process automation support vendor is Electric Water. Online research provides their main office address contact information and key staff. Additional research produces the employee's home addresses.

Online research also provides integrator staff's profiles on LinkedIn. This identified key information on their expertise. Social engineering confirms the main office's Internet service provider, and further reconnaissance provides the router Wi-Fi network ID and credentials.

Waterville Treatment Plant Network

ISP
Internet

Firewall   Server   Firewall   Switch

VPN Connection

Process Control Network VLAN

Process Control Network

Integrator's
Remote Access

Process Control Network
Server/Gateway

Switch

NaOCl Dosage Process

Switch   Workstation/SCADA
Interface   OIT

NaOCl Dosage Process

PLC

NaOCl Dosage Process Equipment

Sensor   Valves   Motor   Other

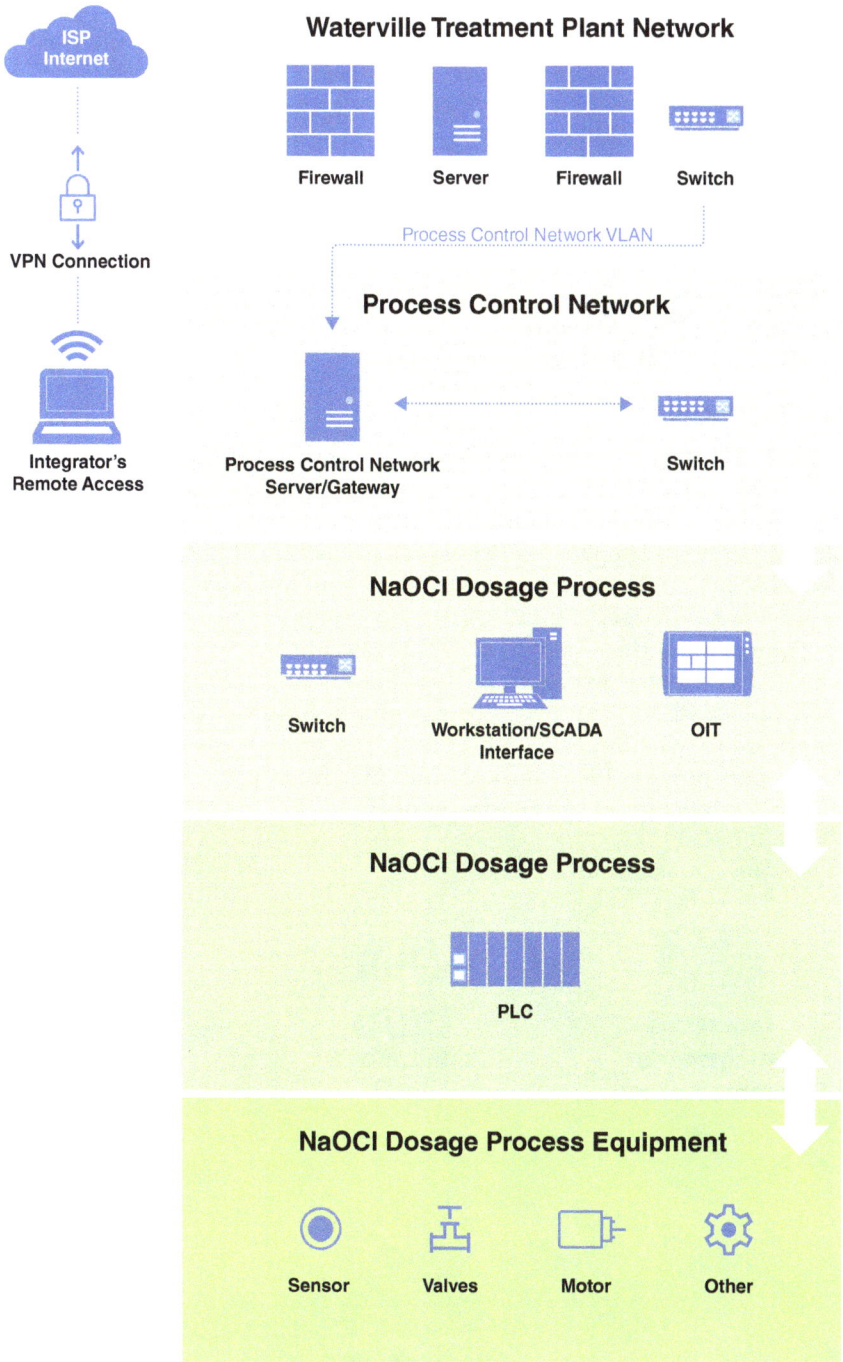

ISP—Internet service provider, OIT—operator interface terminal, PLC—programmable logic circuit, SCADA—supervisory control and data acquisition, VPN—virtual private network

Figure A-15   Network architecture—Integrator's access

Vendor Development/File Workstation

Investigation of available documentation on the development/file host produces a "remote access procedure" for automation support, District SCADA network access credentials, and all of the various technical data related to the design and active maintenance of the control system. Details available on the workstation include the following:

Host. Vendor: Dell; Model: Precision 3630 Tower; OS/Misc.: x64, Windows 10 Enterprise; and Supported Protocols: TCP/IP/Ethernet, SSH, SNMP, HTTPS, HTTP

Software. Vendor: Inductive Automation; Name: Ignition PLC Confirmation Software; Function: SCADA configuration

Technical Approach

Target 1 (T1). Electric water development/file workstation

Access: Leveraging poor network security configuration (e.g., open broadcast of service set identifier, no media access control filter, default router credentials, unpatched firmware) on Electric Water's Wi-Fi router, the adversary can compromise a workstation using the persistent connection to the Wi-Fi network (see Figure A-16).

Timing/triggering: Immediately after access to the company Wi-Fi network and connected workstation.

Action/payload: The malware payload is installed on the workstation. The malware will target, compromise, and modify logic in the District's disinfection process PLC the next time Electric Water establishes a remote network connection to the SCADA system.

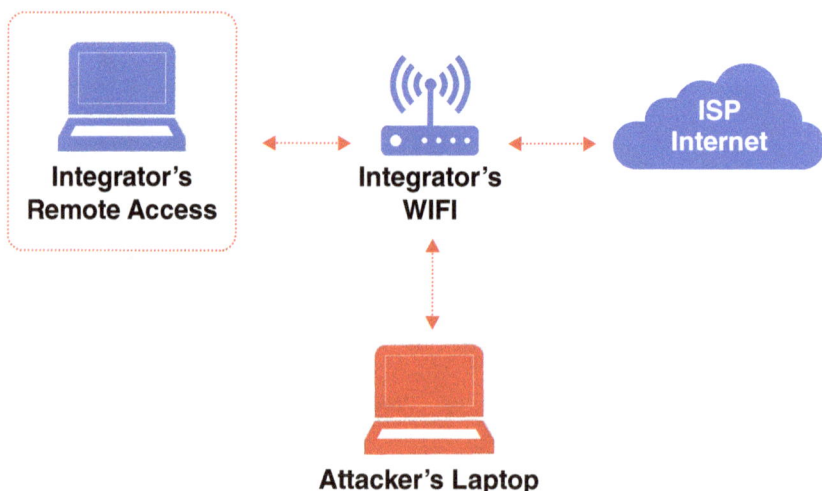

Figure A-16   Access to target 1

Target 2 (T2). Process controller: Modicon 340

Access: Access is established through trusted remote connectivity from Electric Water workstation via certificate-based authentication and VPN server configuration at the District's Internet-facing firewall. Leveraging the SCADA technician's account (escalated privileges)—and the treatment environment's flat network architecture, as well as the absence of secondary independent authentication requirement—Electric Water's workstation quickly establishes a secure session with the disinfection PLC (see Figure A-17).

Timing/triggering: Immediately when Electric Water's VPN establishes a connection to the District's SCADA network.

Action/payload: The action involves changing NaOCl feed set rates in the Modicon PLC to cease any addition of the disinfectant. In the PLC, the adversary will modify proportional–integral–derivative coefficients to produce a targeted (0% dosing) set point and associated tag(s), which is communicated to the chemical feed pump. Supporting operations and instrumentation tags will be modified to static "acceptable range" values before being communicated to the HMI. Tags related to normal chlorine analyzer control communications will also be modified to static "acceptable range" values.

Target Details

The technical approach described in the previous section identifies components that are integral and critical to carrying out the HCE attack. These components provide access, control, and monitoring for the target system.

Adversary efforts to expand understanding of these critical systems and devices would involve development of a critical component list (see Table A-9) providing technical details about each targeted element. Notice that the components list is a subset of taxonomy line items from phase 2. In this case study example, the list would include details available via open-source vendor literature that were not provided already on the compromised workstation.

Critical Needs for Development

To develop the attack that delivers the HCE, an adversary would need to understand the detailed functionality of each critical component, as well as the operational context for use of the technologies. Documentation providing these functional and contextual details would be part of the adversary's critical needs. Table A-10 provides an example critical needs list including likely information location.

Critical Needs for Deployment

The additional element required for payload deployment is access to Electric Water's office Wi-Fi network. Everything else required for deployment is in the payload already as noted in Table A-10.

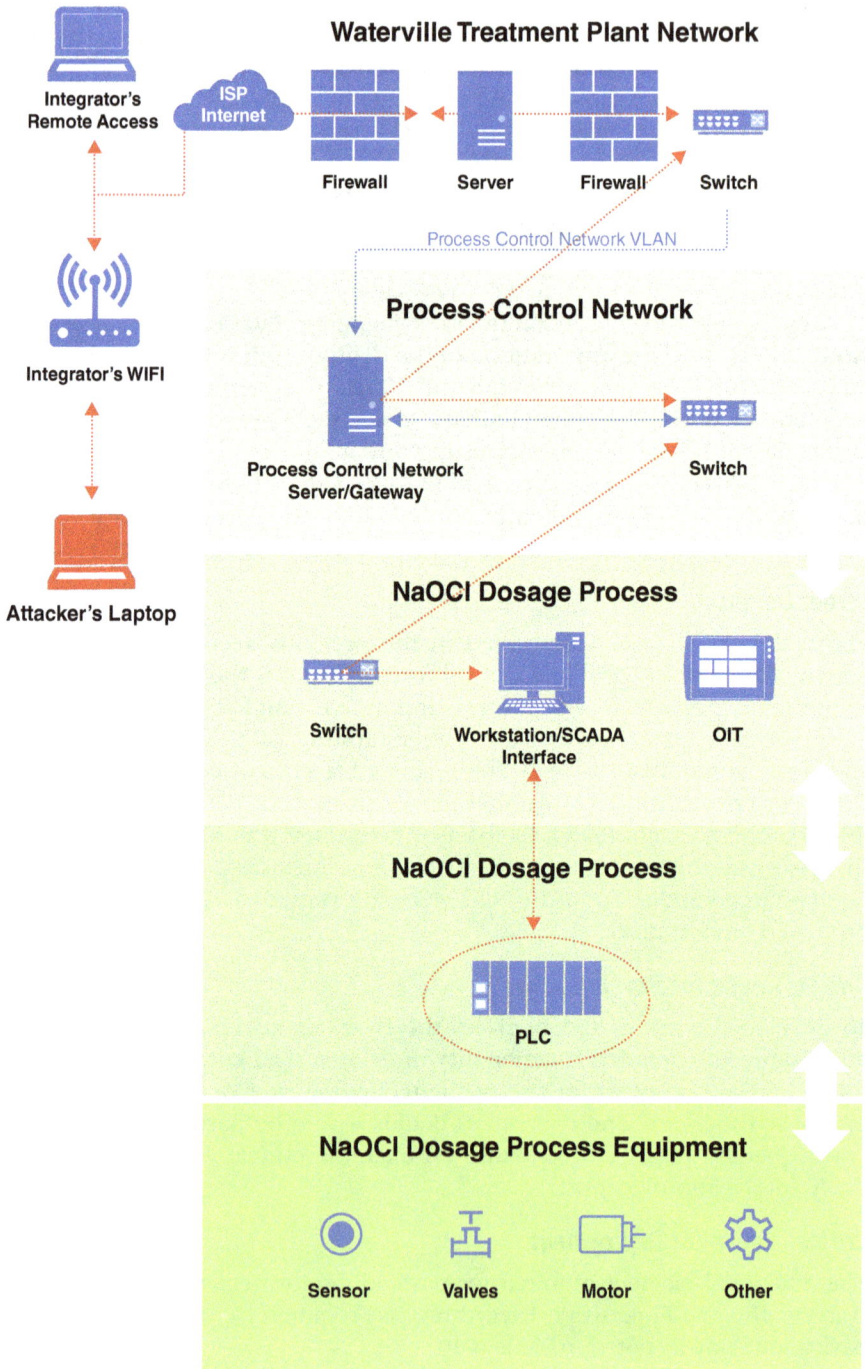

ISP—Internet service provider, OIT—operator interface terminal, PLC—programmable logic circuit, SCADA—supervisory control and data acquisition, VPN—virtual private network

Figure A-17   Access to target 2

Table A-9   Critical components list with target details

Electric Water workstation	Name	Dell Laptop
	Function	SCADA Engineer App/Data Host
	Vendor	Dell
	Model	Precision 3630 Tower
	OS/Misc.	X64, Windows 10 Enterprise
	Protocols	TCP/IP/Ethernet, SSH, SNMP, HTTPS, NTTP
Disinfection process controller	Name	Disinfection Process PLC
	Function	Disinfection process controller
	Vendor	Schneider Electric
	Model	Modicon M340
	OS/Misc.	Version 3.30
	Protocols	Modbus
Disinfection process HMI	Name	Disinfectant process HMI
	Function	Disinfectant process–Operator interface
	Vendor	Allen-Bradley
	Model	PanelView Plus 6
	Protocols	Ethernet, Modbus TCP
Modicon PLC configuration software	Name	Schneider Electric Control Expert
	Function	PLC (M340) configuration
	Vendor	Schneider Electric
Electric Water Wi-Fi router	Name	Electric Water Wi-Fi router
	Function	Home Wi-Fi network router
	Vendor	Cisco Meraki
	Model	MX64
	Protocols	Ethernet

HMI—human–machine interface, PLC—programmable logic circuit

Phase 4: Mitigations and Protections

Phase 4 is Mitigations and Protections. Recall the framework from the lesson plan as shown in Figure A-18. Using this framework, protection and mitigation recommendations are provided to protect and mitigate the HCE.

Protect.
The following protections are identified:
- Ensure hardware protection via fob or card key.
- Install NaOCl flow switch for physical protection on the chemical feed pump.
- Eliminate all remote access to the SCADA system. Any contractors/manufacturers must be on-site to access the environment.
- Install low/low-low alarm safety systems.
- Implement flow monitoring on diffusers in the system. Make independent system where alarms go off. If chlorine stops, no flow and alarm.
- Implement an offline chemical analyzer that hardware enables the chemical feed pumps.

Table A-10   Critical needs

Component	Critical Needs for Development	Location/Availability
Electric Water Wi-Fi router	Vendor specs/data sheets Vendor Ops manual Configuration file	Open source Vendor website Onboard, available at initial compromise
Schneider Electric Modicon M340 PLC	Vendor specs/data sheet Vendor Ops manual Vendor I/O module pinouts/wiring diagrams Waterville PLC project file	Open source Vendor website Vendor website Electric Water workstation; company data/file server
HMI	Vendor specs/data sheets Vendor ops manual Waterville HMI project file	Open source Vendor website Electric Water workstation; company data/file server
PLC configuration	Software and associated install documentation Vendor programming literature	Purchase Vendor website
HMI screen builder	Software and associated install documentation Vender programming literature	Purchase Vendor website
Electric Water workstation	Vendor specs/data sheets Operating system VPN	Open source Open source Onboard, available at compromise

HMI—human–machine interface, I/O—input/output, PLC—programmable logic circuit, VPN—virtual private network

- Implement a policy that only utility-provided computers may connect to the SCADA environment.
- Implement device access control list on network switch(es).

Detect.
If remote connections are authorized,
- implement multifactor authentication;
- enforce alerts for remote connections; and
- implement network intrusion detection/monitoring.

Respond.
- Train and exercise staff on switching to manual control.
- Obtain grab samples to ensure water quality.

Recover.
- Start recovering by manual control and monitoring. Once there is confidence in the safe operation of system through new operating procedures that can confidently verify the process is correctly shown at the SCADA screen, start Auto-mode again.

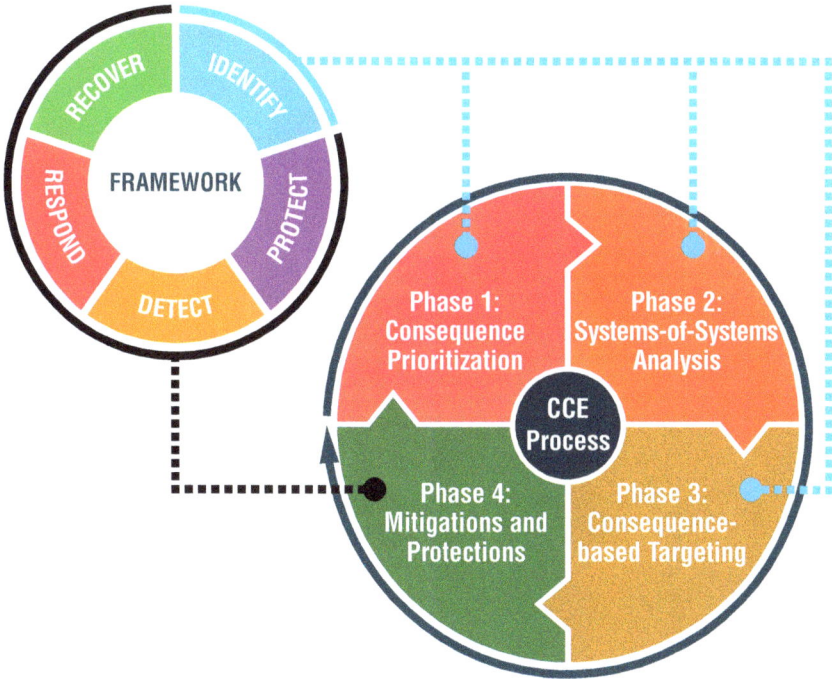

Source: Used with Permission from BEA/INL

Figure A-18   **National Institute of Standards and Technology (NIST) Cybersecurity Framework applied to INL–CCE**

REFERENCES

ATS (American Trauma Society). "Trauma Center Levels Explained." www.amtrauma.org/page/traumalevels (accessed April 15, 2025).

Water Supply Committee of the Great Lakes–Upper Mississippi River Board of State and Provincial Public Health and Environmental Managers. 2012. *Recommended Standards For Water Works: Policies for the Review and Approval of Plans and Specifications for Public Water Supplies.* Albany, N.Y.: Health Research Inc., Health Education Services Division.

Integrating CIE with AWWA's J100 Standard

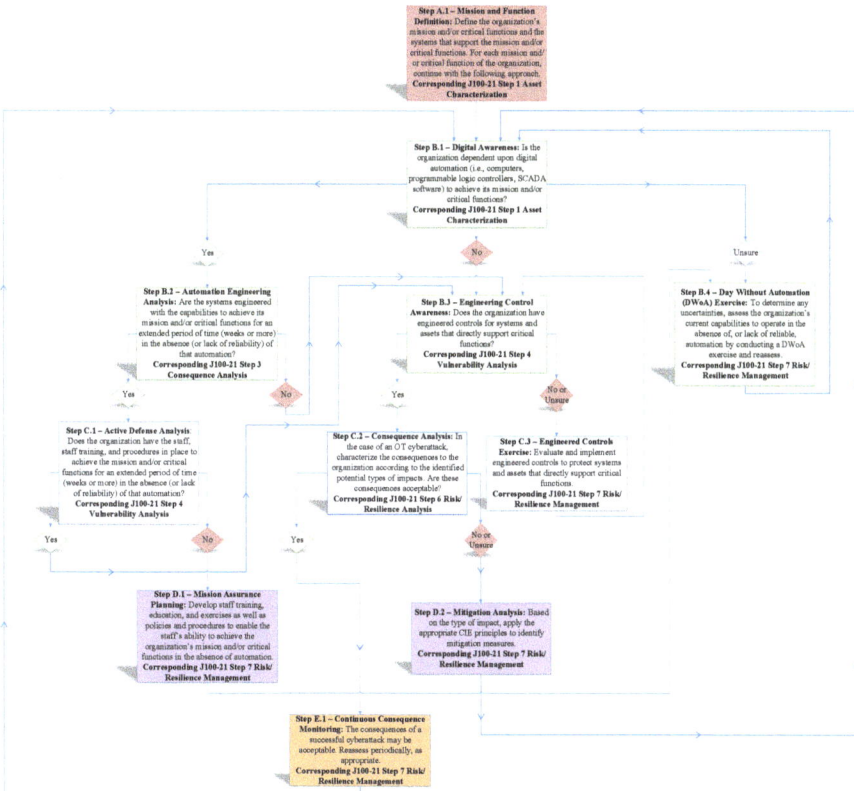

Step A.1 – Mission and Function Definition: Define the organization's mission and/or critical functions and the systems that support the mission and/or critical functions. For each mission and/or critical function of the organization, continue with the following approach. **Corresponding J100-21 Step 1 Asset Characterization**

Step B.1 – Digital Awareness: Is the organization dependent upon digital automation (i.e., computers, programmable logic controllers, SCADA software) to achieve its mission and/or critical functions? **Corresponding J100-21 Step 1 Asset Characterization**

Yes — No — Unsure

Step B.2 – Automation Engineering Analysis: Are the systems engineered with the capabilities to achieve its mission and/or critical functions for an extended period of time (weeks or more) in the absence (or lack of reliability) of that automation? **Corresponding J100-21 Step 3 Consequence Analysis**

Step B.3 – Engineering Control Awareness: Does the organization have engineered controls for systems and assets that directly support critical functions? **Corresponding J100-21 Step 4 Vulnerability Analysis**

Step B.4 – Day Without Automation (DWoA) Exercise: To determine any uncertainties, assess the organization's current capabilities to operate in the absence of, or lack of reliable, automation by conducting a DWoA exercise and reassess. **Corresponding J100-21 Step 7 Risk/ Resilience Management**

Yes — No — Yes — No or Unsure

Step C.1 – Active Defense Analysis: Does the organization have the staff, staff training, and procedures in place to achieve the mission and/or critical functions for an extended period of time (weeks or more) in the absence (or lack of reliability) of that automation? **Corresponding J100-21 Step 4 Vulnerability Analysis**

Step C.2 – Consequence Analysis: In the case of an OT cyberattack, characterize the consequences to the organization according to the identified potential types of impacts. Are these consequences acceptable? **Corresponding J100-21 Step 6 Risk/ Resilience Analysis**

Step C.3 – Engineered Controls Exercise: Evaluate and implement engineered controls to protect systems and assets that directly support critical functions. **Corresponding J100-21 Step 7 Risk/ Resilience Management**

Yes — No — Yes — No or Unsure

Step D.1 – Mission Assurance Planning: Develop staff training, education, and exercises as well as policies and procedures to enable the staff's ability to achieve the organization's mission and/or critical functions in the absence of automation. **Corresponding J100-21 Step 7 Risk/ Resilience Management**

Step D.2 – Mitigation Analysis: Based on the type of impact, apply the appropriate CIE principles to identify mitigation measures. **Corresponding J100-21 Step 7 Risk/ Resilience Management**

Step E.1 – Continuous Consequence Monitoring: The consequences of a successful cyberattack may be acceptable. Reassess periodically, as appropriate. **Corresponding J100-21 Step 7 Risk/ Resilience Management**

105

Appendix B – Flowchart provides a visual guide to continuously evaluating cyber-informed engineering (CIE) measures implemented within a utility. Beginning with step A.1, users must assess each step and make a decision regarding their cyber hygiene. As users work through the flowchart, they will be given exercises and improvements to implement within the utility, eventually ending in continuous consequence monitoring. For further information, readers are encouraged to visit the US Department of Energy's Office of Scientific and Technical Information (OSTI) website. The OSTI website offers comprehensive resources and documentation on CIE, providing users with the most current and reliable information.

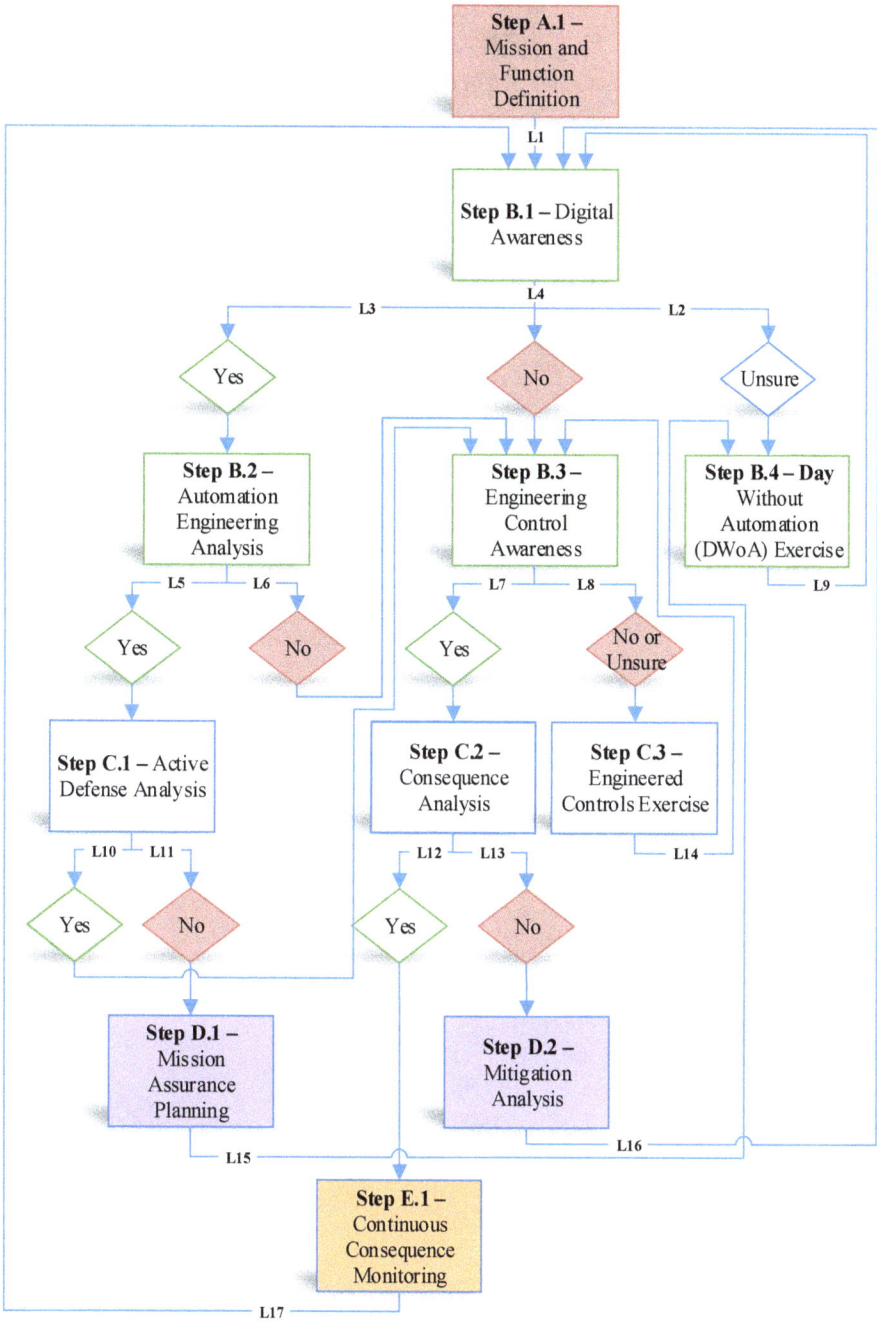

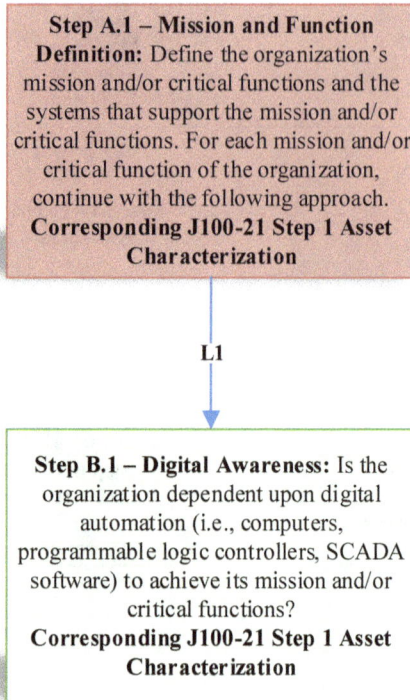

Step A.1 – Mission and Function Definition: Define the organization's mission and/or critical functions and the systems that support the mission and/or critical functions. For each mission and/or critical function of the organization, continue with the following approach. **Corresponding J100-21 Step 1 Asset Characterization**

L1

Step B.1 – Digital Awareness: Is the organization dependent upon digital automation (i.e., computers, programmable logic controllers, SCADA software) to achieve its mission and/or critical functions? **Corresponding J100-21 Step 1 Asset Characterization**

Step B.1 – Digital Awareness: Is the organization dependent upon digital automation (i.e., computers, programmable logic controllers, SCADA software) to achieve its mission and/or critical functions?
Corresponding J100-21 Step 1 Asset Characterization

L2

Unsure

Step B.4 – Day Without Automation (DWoA) Exercise: To determine any uncertainties, assess the organization's current capabilities to operate in the absence of, or lack of reliable, automation by conducting a DWoA exercise and reassess.
Corresponding J100-21 Step 7 Risk/ Resilience Management

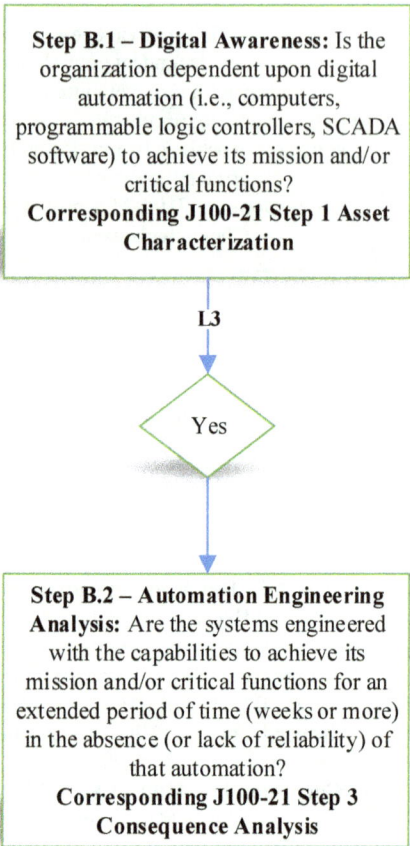

Step B.1 – Digital Awareness: Is the organization dependent upon digital automation (i.e., computers, programmable logic controllers, SCADA software) to achieve its mission and/or critical functions?
Corresponding J100-21 Step 1 Asset Characterization

L4

No

Step B.3 – Engineering Control Awareness: Does the organization have engineered controls for systems and assets that directly support critical functions?
Corresponding J100-21 Step 4 Vulnerability Analysis

Step B.2 – Automation Engineering Analysis: Are the systems engineered with the capabilities to achieve its mission and/or critical functions for an extended period of time (weeks or more) in the absence (or lack of reliability) of that automation?
Corresponding J100-21 Step 3 Consequence Analysis

L5

Yes

Step C.1 – Active Defense Analysis: Does the organization have the staff, staff training, and procedures in place to achieve the mission and/or critical functions for an extended period of time (weeks or more) in the absence (or lack of reliability) of that automation?
Corresponding J100-21 Step 4 Vulnerability Analysis

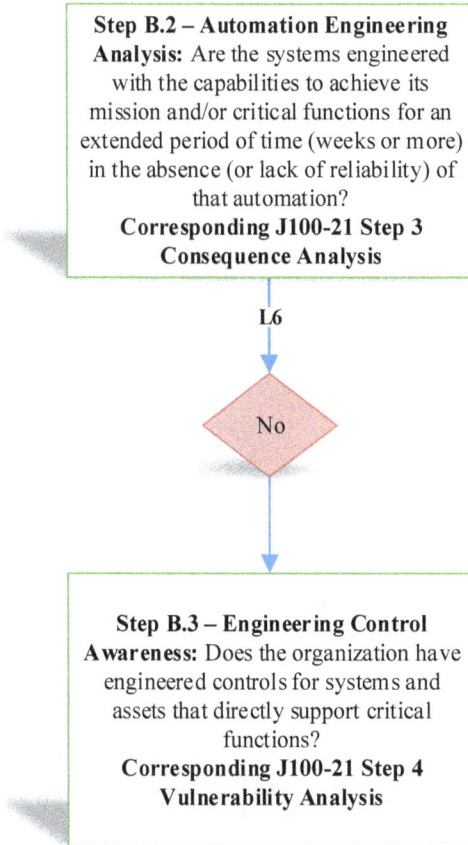

Step B.3 – Engineering Control Awareness: Does the organization have engineered controls for systems and assets that directly support critical functions?
Corresponding J100-21 Step 4 Vulnerability Analysis

L7

Yes

Step C.2 – Consequence Analysis: In the case of an OT cyberattack, characterize the consequences to the organization according to the identified potential types of impacts. Are these consequences acceptable?
Corresponding J100-21 Step 6 Risk/ Resilience Analysis

Step B.3 – Engineering Control Awareness: Does the organization have engineered controls for systems and assets that directly support critical functions?
Corresponding J100-21 Step 4 Vulnerability Analysis

L8

No or Unsure

Step C.3 – Engineered Controls Exercise: Evaluate and implement engineered controls to protect systems and assets that directly support critical functions.
Corresponding J100-21 Step 7 Risk/ Resilience Management

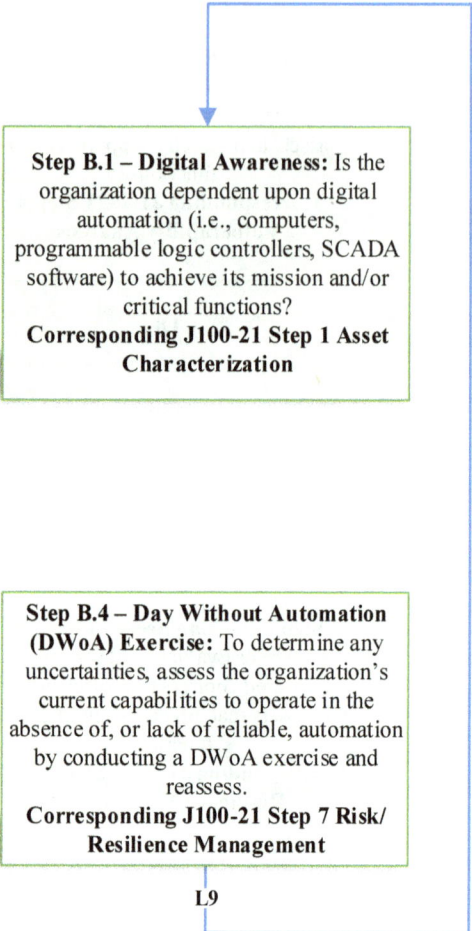

Step B.1 – Digital Awareness: Is the organization dependent upon digital automation (i.e., computers, programmable logic controllers, SCADA software) to achieve its mission and/or critical functions?
Corresponding J100-21 Step 1 Asset Characterization

Step B.4 – Day Without Automation (DWoA) Exercise: To determine any uncertainties, assess the organization's current capabilities to operate in the absence of, or lack of reliable, automation by conducting a DWoA exercise and reassess.
Corresponding J100-21 Step 7 Risk/ Resilience Management

L9

Step C.1 – Active Defense Analysis:
Does the organization have the staff, staff training, and procedures in place to achieve the mission and/or critical functions for an extended period of time (weeks or more) in the absence (or lack of reliability) of that automation?
Corresponding J100-21 Step 4 Vulnerability Analysis

L10

Yes

Step B.3 – Engineering Control Awareness: Does the organization have engineered controls for systems and assets that directly support critical functions?
Corresponding J100-21 Step 4 Vulnerability Analysis

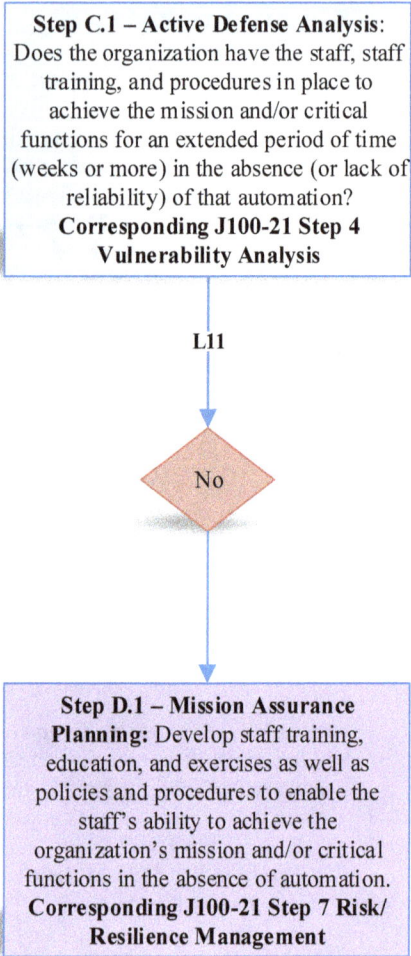

Step C.1 – Active Defense Analysis: Does the organization have the staff, staff training, and procedures in place to achieve the mission and/or critical functions for an extended period of time (weeks or more) in the absence (or lack of reliability) of that automation? **Corresponding J100-21 Step 4 Vulnerability Analysis**

L11

No

Step D.1 – Mission Assurance Planning: Develop staff training, education, and exercises as well as policies and procedures to enable the staff's ability to achieve the organization's mission and/or critical functions in the absence of automation. **Corresponding J100-21 Step 7 Risk/ Resilience Management**

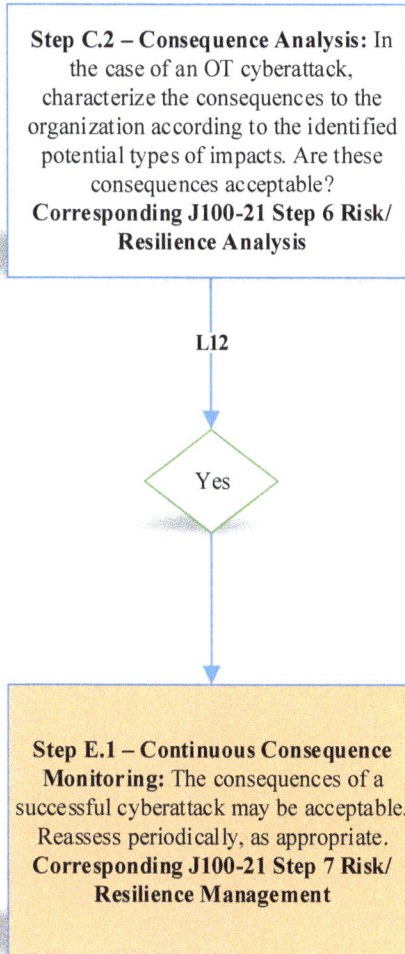

Step C.2 – Consequence Analysis: In the case of an OT cyberattack, characterize the consequences to the organization according to the identified potential types of impacts. Are these consequences acceptable?
Corresponding J100-21 Step 6 Risk/ Resilience Analysis

L12

Yes

Step E.1 – Continuous Consequence Monitoring: The consequences of a successful cyberattack may be acceptable. Reassess periodically, as appropriate.
Corresponding J100-21 Step 7 Risk/ Resilience Management

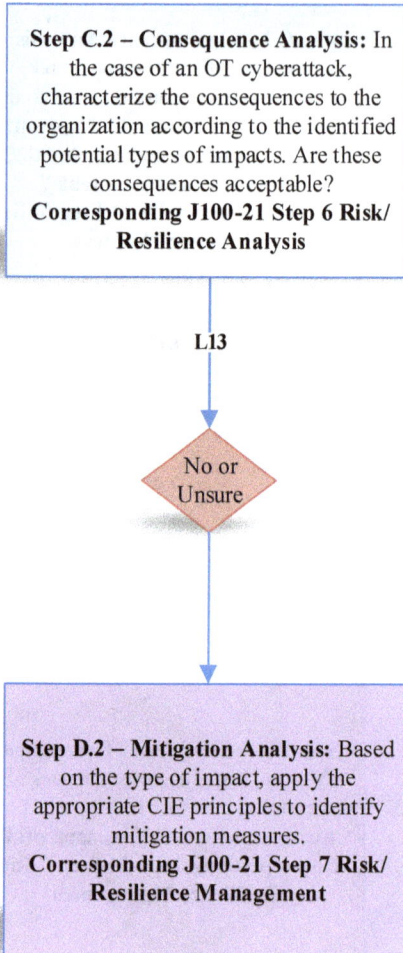

Step B.3 – Engineering Control Awareness: Does the organization have engineered controls for systems and assets that directly support critical functions?
Corresponding J100-21 Step 4 Vulnerability Analysis

Step C.3 – Engineered Controls Exercise: Evaluate and implement engineered controls to protect systems and assets that directly support critical functions.
Corresponding J100-21 Step 7 Risk/ Resilience Management

L14

Step B.4 – Day Without Automation (DWoA) Exercise: To determine any uncertainties, assess the organization's current capabilities to operate in the absence of, or lack of reliable, automation by conducting a DWoA exercise and reassess.
Corresponding J100-21 Step 7 Risk/ Resilience Management

Step D.1 – Mission Assurance Planning: Develop staff training, education, and exercises as well as policies and procedures to enable the staff's ability to achieve the organization's mission and/or critical functions in the absence of automation.
Corresponding J100-21 Step 7 Risk/ Resilience Management

L15

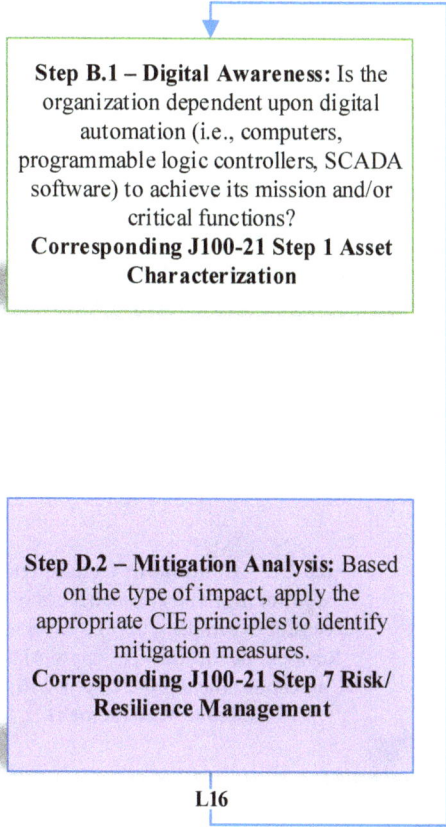

Step B.1 – Digital Awareness: Is the organization dependent upon digital automation (i.e., computers, programmable logic controllers, SCADA software) to achieve its mission and/or critical functions?
Corresponding J100-21 Step 1 Asset Characterization

Step D.2 – Mitigation Analysis: Based on the type of impact, apply the appropriate CIE principles to identify mitigation measures.
Corresponding J100-21 Step 7 Risk/ Resilience Management

L16

Step B.1 – Digital Awareness: Is the organization dependent upon digital automation (i.e., computers, programmable logic controllers, SCADA software) to achieve its mission and/or critical functions?
Corresponding J100-21 Step 1 Asset Characterization

Step E.1 – Continuous Consequence Monitoring: The consequences of a successful cyberattack may be acceptable. Reassess periodically, as appropriate.
Corresponding J100-21 Step 7 Risk/ Resilience Management

L17

www.ingramcontent.com/pod-product-compliance
Lightning Source LLC
Chambersburg PA
CBHW070729220326
41598CB00024BA/3370